Ecology and conservation / Écologie et conservation 6

Titles in this series / Dans cette collection:

1. *Ecology of the subarctic regions. Proceedings of the Helsinki symposium / Écologie des régions subarctiques. Actes du colloque d'Helsinki.*
2. *Methods of study in soil ecology. Proceedings of the Paris symposium / Méthodes d'étude de l'écologie des sols. Actes du colloque de Paris.*
3. *The origin of* Homo sapiens. *Proceedings of the Paris symposium / Origine de l'homme moderne. Actes du colloque de Paris.*
4. *Productivity of forest ecosystems. Proceedings of the Brussels Symposium / La productivité des écosystèmes forestiers. Actes du colloque de Bruxelles.*
5. *Plant response to climatic factors. Proceedings of the Uppsala Symposium / Réponse des plantes aux facteurs climatiques. Actes du colloque d'Uppsala.*
6. *International classification and mapping of vegetation / Classification internationale et cartographie de la végétation / Clasificación internacional y cartografía de la vegetación.*

International classification and mapping of vegetation

Classification internationale et cartographie de la végétation

Clasificación internacional y cartografía de la vegetación

Unesco Paris 1973

Published by the United Nations Educational,
Scientific and Cultural Organization

Publié par l'Organisation des Nations Unies
pour l'éducation, la science et la culture

Publicado por la Organización de las Naciones Unidas
para la Educación, la Ciencia y la Cultura

7, place de Fontenoy, 75700 Paris

Printed by / Imprimé par / Impreso por
Imprimeries Populaires de Genève

ISBN 92-3-001046-4
LC No. 72-96442

Preface

A first list of concepts and symbols prepared by Dr Schmitthüsen, Professor at the Geographical Institute of Saarbrücken, and Professor Ellenberg, of the Institute of Systematic Phytogeography in the University of Göttingen (Federal Republic of Germany), issued in 1964 and later revised by Dr Poore, Director of Nature Conservancy (United Kingdom), and Professor Ellenberg in 1965, was discussed on several occasions by the Unesco Standing Committee on Classification and Mapping of Vegetation on a World Basis, in the light of preliminary comments, more particularly at Unesco Headquarters in January 1966 and June 1967. The committee was composed of the following members: H. Gaussen (President), Professor at the University of Toulouse and Director of the Institute for the International Mapping of Vegetation Cover (France), R. Germain (Belgium), A. W. Küchler, Professor at the University of Kansas (United States of America), J. Lebrun, Professor at the University of Louvain (Belgium), D. Poore, V. Sočava, of the Botanical Institute of Leningrad (U.S.S.R.), C. Troll, of the Geographical Institute in the University of Bonn (Federal Republic of Germany), and the following Unesco experts: G. Budowski, S. Evteev and O. Fränzle.

The present classification is the outcome of the final revisions made by the above committee in Seattle (United States) in August 1969, using as a basis for discussion the comments and criticisms received in connexion with the draft classification prepared by Professor Ellenberg and Dr Müller-Dombois, Associate Professor of Botany at the University of Hawaii, at the committee's request, which was published and widely distributed in the *Ber. geobot. Inst. ETH Stiftg Rübel, Zürich, 1965-66* (Vol. 37, 1967, p. 21-46) under the title: "Tentative Physiognomic-ecological Classification and Mapping".

The "Framework for a Classification of World Vegetation" (the Unesco classification) was tested in 1970 under tropical field conditions in Costa Rica by A. W. Küchler and J. M. Montoya-Maquin.[1] The tests revealed that the Unesco classification of vegetation could be applied with relative ease and accuracy, requiring only minor adjustments which have since been made.

The exception to the rule was the part of the classification devoted to herbaceous vegetation. Professor Küchler then revised this part, and submitted the revision to the members of the Unesco Standing Committee on Classification and Mapping of Vegetation. It was also submitted to a number of particularly qualified scientists outside the committee. The revision was overwhelmingly approved. The approval was sometimes linked with various suggestions for improvements and refinements, and some of these were not limited to the sections on herbaceous vegetation. All suggestions have been integrated into the classification to the limit of feasibility.

Unesco is particularly grateful to Professor H. Gaussen, president of the committee and chief editor of the published text of this technical paper, Professor H. Ellenberg, co-author of the drafts

1. "The Unesco Classification of Vegetation: Some Tests in the Tropics. Instituto Inter-Americano de Ciencias Agrícolas", *Turrialba* (Costa Rica), Vol. 21, 1971, p. 98-109.

which formed the basis for the discussions, Professor A. W. Küchler for his very active share in putting the finishing touches to the classification, and also to Drs Schmitthüsen, Poore and Müller-Dombois who took part in the preparation of the basic drafts. Unesco also expresses its gratitude to J. M. Montoya-Maquin, of the OAS Inter-American Institute of Agricultural Sciences (Costa Rica), and H. Jiménez-Saa, of the Centre for Post-graduate Studies in Forestry at the University of the Andes (Venezuela) as well as to the numerous experts who have taken part in the preparation of this classification, and especially to all those who have sent written comments and criticisms.

Furthermore, F. Bagnouls and P. Legris are congratulated on the choice of colours made in collaboration with Professor Gaussen on the basis of the general principles published by him in 1926.

Finally, the presentation of the colour scale required delicate workmanship, for which Unesco is grateful to the National Geographical Institute of Paris.

Préface

Une première liste de concepts et de symboles, préparée en 1964 par MM. Schmitthüsen, professeur à l'Institut de géographie de Sarrebruck, et Ellenberg, professeur à l'Institut de géobotanique systématique de l'Université de Göttingen (République fédérale d'Allemagne), révisée ensuite en 1965 par MM. Poore, directeur de Nature Conservancy (Royaume-Uni), et Ellenberg a été discutée à plusieurs reprises par le comité permanent établi par l'Unesco pour la classification et la cartographie de la végétation du monde, à la lumière des premiers commentaires, notamment au siège de l'Unesco en janvier 1966 et en juin 1967. Les membres de ce comité étaient les suivants: H. Gaussen (président), professeur à l'Université de Toulouse et directeur de l'Institut de la « Carte internationale du tapis végétal »; R. Germain, du Laboratoire d'écologie végétale de l'Université de Louvain (Belgique); A. W. Küchler, professeur à l'Université du Kansas (États-Unis d'Amérique); J. Lebrun, professeur à l'Université de Louvain (Belgique); D. Poore, V. Sočava, de l'Institut de botanique de Leningrad (URSS); C. Troll, de l'Institut de géographie de l'Université de Bonn (République fédérale d'Allemagne); ainsi que les spécialistes de l'Unesco G. Budowski, S. Evteev et O. Fränzle.

La présente classification est le résultat des dernières révisions faites par ce comité à Seattle, Wash. (États-Unis d'Amérique), en août 1969, sur la base des commentaires et critiques dont a fait l'objet le projet préparé par H. Ellenberg et D. Müller-Dombois, professeur adjoint à l'Université de Hawaii, à la demande du comité. Ce projet de classification a été publié et largement distribué dans le *Ber. Geob. Inst. ETH Stiftg Rübel, Zurich, 1965-1966* (vol. 37, p. 21-46, 1967) sous le titre « Tentative physiognomic-ecological classification and mapping » (Essai de classification écologico-physionomique des formations végétales du monde).

La « classification Unesco » (Framework for a classification of world vegetation) a donné lieu en 1970 à des essais d'application dans la zone tropicale au Costa Rica par A. W. Küchler et J. M. Montoya Maquin [1]. Ces essais ont montré que la classification de la végétation préconisée par l'Unesco était relativement facile à appliquer et suffisamment précise et ne nécessitait que des ajustements mineurs qui ont été faits par la suite.

Une seule exception: la partie consacrée à la végétation herbacée. Le professeur Küchler a revu cette partie et a soumis son travail aux membres du Comité permanent de l'Unesco pour la classification et la cartographie de la végétation. Cette révision a également été présentée à un certain nombre de spécialistes particulièrement qualifiés ne faisant pas partie du comité. Elle a été très largement approuvée. Les approbations ont été quelquefois accompagnées de diverses suggestions tendant à apporter des améliorations et des perfectionnements dont certains n'étaient pas limités aux sections portant sur la végétation herbacée. Dans la mesure du possible toutes les suggestions ont été incorporées à la classification.

1. "The Unesco classification of vegetation: some tests in the tropics", *Turrialba*, vol. 21, p. 98-109, Turrialba (Costa Rica), Instituto Interamericano de Ciencias Agrícolas, 1971.

L'Unesco est particulièrement reconnaissante au professeur H. Gaussen, président du comité et principal auteur du texte de présentation de cette note technique, au professeur H. Ellenberg, l'un des rédacteurs des projets qui ont servi de base aux débats, au professeur A. W. Küchler pour sa participation particulièrement active et précieuse, et à MM. Schmitthüsen, Poore et Müller-Dombois, qui ont pris part à la rédaction des projets de base. L'Unesco exprime sa gratitude à MM. Montoya-Maquin, de l'Institut interaméricain des sciences agricoles de l'Organisation des États américains (Costa Rica), et H. Jiménez-Saa, du Centre de formation supérieure de l'Université des Andes (Venezuela), ainsi qu'aux nombreux spécialistes qui ont participé à la mise au point de la présente classification, en particulier à ceux qui ont envoyé par écrit leurs observations ou critiques.

Doivent également être vivement remerciés MM. Bagnouls et Legris pour le choix des couleurs, qui a été effectué en collaboration avec le professeur Gaussen sur la base des principes généraux publiés par ce dernier en 1926.

D'autre part, la présentation de la gamme des couleurs a nécessité un travail délicat dont l'Unesco est reconnaissante à l'Institut géographique national de Paris.

Prefacio

Una primera lista de conceptos y de símbolos preparada por los Sres. Schmitthüsen, profesor en el Instituto de Geografía de Sarrebruck (República Federal de Alemania), y Ellenberg, profesor del Instituto de Geobotánica Sistemática de la Universidad de Göttingen (República Federal de Alemania), editada en 1964, revisada después en 1965 por los Sres. Poore, director de Nature Conservancy (Reino Unido) y Ellenberg, fue examinada en varias ocasiones por el Comité Permanente de la Unesco para la Clasificación y la Cartografía de la Vegetación sobre una Base Mundial, a la vista de los primeros comentarios, especialmente en la Sede de la Unesco en enero de 1966 y en junio de 1967. Los miembros de ese comité eran los siguientes: H. Gaussen (presidente), profesor de la Universidad de Toulouse y director del Instituto del Mapa Internacional del Tapiz Vegetal; R. Germain (Bélgica); A. W. Küchler, profesor de la Universidad de Kansas (Estados Unidos de América); J. Lebrun, profesor de la Universidad de Lovaina (Bélgica); D. Poore, V. Sočava, del Instituto de Botánica de Leningrado (URSS); C. Troll, del Instituto de Geografía de la Universidad de Bonn (República Federal de Alemania); así como G. Budowski, S. Evteev y O. Fränzle, especialistas de la Unesco.

La actual clasificación es el resultado de las últimas revisiones efectuadas por este comité en Seattle, Wash. (Estados Unidos de América), en agosto de 1969, basándose en los comentarios y críticas recibidos sobre el proyecto preparado por H. Ellenberg y D. Müller-Dombois, profesor asociado de la Universidad de Hawaii, a petición del comité. En el *Ber. Geob. Inst. ETH Stiftg Rubel, Zurich 1965-1966* (vol. 37, p. 21-46, 1967), se ha publicado y distribuido ampliamente este proyecto de clasificación con el título de «Tentative physiognomic-ecological classification and mapping» (Ensayo de clasificación ecológico-fisionómica de las formaciones vegetales del mundo).

La "clasificación de la Unesco" ("Framework for a classification of world vegetation") fue ensayada en 1970 en las condiciones tropicales locales de Costa Rica por A. W. Küchler y J. M. Montoya Maquin [1]. Los ensayos demostraron que la clasificación de la vegetación de la Unesco podía aplicarse con relativa facilidad y exactitud y sólo requería pequeños reajustes que después se han efectuado.

La excepción a la regla fue la parte de la clasificación dedicada a la vegetación herbácea. El profesor Küchler revisó esta parte y sometió la revisión a los miembros del Comité Permanente de la Unesco para la Clasificación y la Cartografía de la Vegetación sobre una Base Mundial. También se sometió a algunos científicos muy calificados ajenos al comité. La revisión fue aprobada casi sin objeciones. En algunos casos, la aprobación iba acompañada de diversas sugestiones de mejoramientos y refinamientos, algunos de los cuales no se limitaban a las secciones sobre vegetación herbácea. Todas las sugestiones se han incluido en la clasificación dentro del límite de las posibilidades.

La Unesco está particularmente agradecida al profesor H. Gaussen, presidente del Comité y

1. "The Unesco classification of vegetation: some tests in the tropics", *Turrialba*, vol. 21, p. 98-108, Turrialba (Costa Rica), Instituto Interamericano de Ciencias Agrícolas, 1971.

principal autor del texto de presentación de esta nota técnica, al profesor H. Ellenberg, que ha sido uno de los autores de los textos utilizados como base de las discusiones, al profesor A. W. Küchler por su activa participación en la redacción final de la clasificación, así como a los Sres. Schmitthüsen, Poore y Müller-Dombois, que participaron en la preparación de los proyectos básicos. Por último, la Unesco expresa su gratitud a los Sres. J. M. Montoya-Maquin, del Instituto Interamericano de Ciencias Agrícolas de la Organización de los Estados Americanos (Costa Rica), y H. Jiménez-Saa, del Centro de Estudios Forestales Postuniversitarios de la Universidad de los Andes (Venezuela), así como a los numerosos especialistas que han participado en la preparación de esta clasificación, y particularmente a todos los que se han dignado enviar comentarios y críticas por escrito.

También se debe agradecer a los Sres. Bagnouls y Legris por la selección de colores hecha en colaboración con el profesor Gaussen a base de los principios generales publicados por él en 1926.

Además, la presentación de la gama de colores ha exigido un trabajo delicado por el que la Unesco está muy agradecida al Institut géographique national de Paris.

Contents

Sommaire

Índice

International classification and mapping of vegetation

Contents

Introduction

The following classification has been prepared by the Unesco Standing Committee on Classification and Mapping of Vegetation on a World Basis. It is to provide a comprehensive framework for the more important categories to be used in vegetation maps at scales of 1 : 1 million or smaller.[1] While the classification is intended primarily for the preparation of new maps, it may also be used for transforming existing vegetation maps into this system.

The committee reviewed existing classifications and concluded that none of them was entirely appropriate for the present purpose, although they have influenced the thinking and decisions of the committee. As a classification the system is necessarily artificial. Ecological and sociological relationships are not implied by the arrangement of the units.

The categories in this classification are units of vegetation, including both zonal formations and the more important and extensive azonal and modified formations. It is difficult to use floristics as a general basis for a world classification, because equivalent formations in different parts of the world are drawn from different floras, although in any particular area floristics usually play a significant part in the definition and distinction of communities.

The best set of variables for world-wide comparison are physiognomy and structure of vegetation. Unfortunately, these two attributes of vegetation are not always clearly identifiable with important ecological habitats or environments. For this reason supplementary terms referring to climate, soil and landforms are included in the names and occasionally in the definitions, where they help in the identification of a given unit. In the preparation of the classification, therefore, it was deemed to be pragmatic rather than strictly systematic, in order to provide units with names and definitions which are short, meaningful and descriptive.

Although most units are defined physiognomically, they broadly indicate environmental conditions as well. The physiognomy of many grasslands does not permit a mapper to recognize their latitudinal affinities. To indicate such affinities in Formation Class V of this physiognomic classification would therefore lead to an undue amount of duplication and repetition. However, authors are urged to employ terms like tropical, temperate, etc. in connexion with herbaceous plant communities when they prepare their map legends, and whenever such supplementary information will help the reader of the vegetation map in identifying the categories involved.

One of the most persistently raised questions about the Unesco classification of vegetation concerns the dynamic aspects of vegetation and the place of the various categories in the succession toward the climax. Obviously, not all categories represent climax conditions. Here again, as in the use of supplementary environmental information, the committee was guided by pragmatic principles. At the scale of 1 : 1,000,000 or less, it is not feasible to show the details that are often required to identify climax communities. Especially with regard to

1. The system may also be applied to larger scales by expanding it through further subdivisions. A few finer subdivisions are shown to serve as samples.

herbaceous communities, the interpretation of the vegetation in terms of climax considerations is often extremely difficult or impossible. As a result, the Unesco classification is based on climax vegetation types wherever this is practical. In addition, near climax conditions and "semi-natural" vegetation types as they exist at the time of observation are also included. This allows the mapper flexibility and permits him to show many types of vegetation the successional status of which is uncertain, open to various interpretations or at present impossible to determine. Cultural vegetation *sensu stricto*, the so-called messicol vegetation[1] is excluded.[2] The readers of vegetation maps based on the Unesco classification of vegetation should not expect to find wheat fields, vineyards, banana plantations or rice paddies. Such items are outside the framework of this classification.

The Unesco classification of vegetation can therefore be described as being basically physiognomic-structural in character with supplementary ecological information integrated into its various categories and applicable to natural and semi-natural vegetation. While relatively complete, it is so organized as to permit the addition of further categories whenever this is desirable.

1. A. W. Küchler, "Natural and Cultural Vegetation", *Professional Geographer*, Vol. 21, 1969, p. 383-5.
2. *Note by Professor Gaussen*. It is quite clear that it would be difficult to show the climax for the cultivated parts, for it would often be forest for example; but cultivated areas occupy a considerable proportion of the earth's surface, and for mapping purposes must be taken into consideration. If we confine ourselves purely to physiognomic characteristics, then cereals will be grasses, beetroots and potatoes forbs and vines and fruit-trees frutices. In the annex, page 37, the method of representing cultivated areas graphically will be indicated.

Classification

In the Unesco classification, units of unequal rank are distinguished from one another by letters and numbers as follows:

I, II, etc. = FORMATION CLASS
A, B, etc. = *FORMATION SUBCLASS*
1, 2, etc. = FORMATION GROUP
a, b, etc. = *Formation*
(1), (2), etc. = Subformation
(a), (b), etc. = Further subdivisions.

I CLOSED FOREST

Formed by trees at least 5 m tall with their crowns interlocking.[1]

I.A *MAINLY EVERGREEN FOREST*, i.e. the canopy is never without green foliage. However, individual trees may shed their leaves.

I.A.1 TROPICAL OMBROPHILOUS FOREST. (Conventionally called tropical rain forest.) Consisting mainly of broad-leaved evergreen trees, neither cold nor drought resistant. Truly evergreen, i.e. the forest canopy remains green all year though a few individual trees may be leafless for a few weeks. Leaves of many species with "drip tips".

1 I.A.1a *Tropical ombrophilous lowland forest.* Composed usually of numerous species of fast-growing trees, many of them exceeding 50 m in height,[2] generally with smooth, often thin bark, some with buttresses. Emergent trees or at least a very uneven canopy often present. Very sparse undergrowth, and this composed mainly of tree reproduction. Palms and other tuft trees usually rare, lianas nearly absent except pseudo-lianas (i.e. plants originating on tree branches, subsequently rooting in the ground, or vice versa). Crustose lichens and green algae are the only constantly present epiphytic life forms; vascular epiphytes are usually not abundant; abundant only in extremely humid situations, e.g. Sumatra, Atrato Valley (Colombia), etc.

1. In reproductive stage or as immature secondary growth temporarily less than 5 m tall, but individuals of scapose life form, i.e. real trees, not shrubs. In subpolar conditions, the limit may be only 3 m, in tropical ones 8 or 10 m.
2. Height limits are only a generalized guide, not an absolute criterion.

2	I.A.1b	*Tropical ombrophilous submontane forest.* Emergent trees largely absent and canopy relatively even. In the undergrowth, forbs common. Vascular epiphytes and pseudolianas abundant, e.g. Atlantic slopes of Costa Rica.
	I.A.1c	*Tropical ombrophilous montane forest.* Abundant vascular and other epiphytes. Tree sizes usually less than 50 m; crowns extending relatively far down the stem. Bark often more or less rough. Undergrowth abundant, often represented by rosulate nano- and microphanerophytes (e.g. tree ferns or small palms); the ground layer rich in hygromorphous herbs and cryptogams, e.g. Sierra de Talamanca, Costa Rica.
3	I.A.1c(1)	Broad-leaved.
4	I.A.1c(2)	Needle-leaved.
5	I.A.1c(3)	Microphyllous.
6	I.A.1c(4)	Bamboo, rich in tree-grasses replacing largely the tuft micro- or nanophanerophytes.
7	I.A.1d	*Tropical ombrophilous "subalpine" forest.* (Excluding cloud forest or woodland. Considered unique by some investigators, but probably not important. Definition required.)
	I.A.1e	*Tropical ombrophilous cloud forest.* Tree crowns, branches and trunks as well as lianas burdened with epiphytes, mainly chamaephytic bryophytes. Also the ground covered with hygromorphic chamaephytes (e.g. *Selaginella* and herbaceous ferns). Trees often gnarled, with rough bark and rarely exceeding 20 m in height, e.g. Blue Mountains, Jamaica.
8	I.A.1e(1)	Broad-leaved, most common form.
9	I.A.1e(2)	Needle-leaved.
10	I.A.1e(3)	Microphyllous.
	I.A.1f	*Tropical ombrophilous alluvial forest.* Similar to submontane forest (I.A.1b), but richer in palms and in undergrowth life forms, particularly tall forbs (e.g. Musaceae); buttresses frequent, e.g. Amazon Basin.
11	I.A.1f (1)	Riparian forest on low, frequently flooded, river banks, mostly dominated by fast-growing trees; herbaceous undergrowth nearly absent; epiphytes rare, poor in species, e.g. the Amazonian igapó.
12	I.A.1f (2)	Occasionally flooded on relatively dry terraces accompanying active rivers. More epiphytes than in (1) and (3), many lianas.
13	I.A.1f (3)	Seasonally water-logged along the lower river courses, where the water accumulates for several months on large flats, often behind low natural levees; trees frequently with stilt roots; canopy density not uniform; as a rule poor in undergrowth, except in more open places, e.g. the Amazonian várzea.
	I.A.1g	*Tropical ombrophilous swamp forest.* Not along rivers, but on edaphically wet habitats, which may be supplied with either fresh or brackish water. Similar to alluvial forests (I.A.1f), but as a rule poorer in tree species. Many trees with buttresses or pneumatophores; mostly taller than 20 m, e.g. in eastern Sumatra.
14	I.A.1g(1)	Broad-leaved, dominated by dicotyledonous plants.
15	I.A.1g(2)	Dominated by palms, but broad-leaved trees in the undergrowth, e.g. the *Raphia taedigera* swamps of Costa Rica.
16	I.A.1h	*Tropical evergreen bog forest* (with organic surface deposits). Poor in tree species, with canopy often forming a pattern of tall trees at the bog fringe to shorter trees near the centre. Trees often have thin diameters and are commonly equipped with pneumatophores or stilt roots.
	I.A.2	TROPICAL AND SUBTROPICAL EVERGREEN SEASONAL FOREST. Consisting mainly of broad-leaved evergreen trees. Foliage reduction during the dry season is

noticeable, often as partial shedding. Transitional between I.A.1 and I.A.3. Subdivisions a-c largely similar to those under I.A.1, the tropical ombrophilous forest.

17 I.A.2a *Tropical or subtropical evergreen seasonal lowland forest.*

I.A.2b *Tropical or subtropical evergreen seasonal submontane forest.*
18 I.A.2b(1) Broad-leaved.
19 I.A.2b(2) Needle-leaved.

20 I.A.2c *Tropical or subtropical evergreen seasonal montane forest.* In contrast to I.A.1c no tree ferns; instead, evergreen shrubs are most frequent.

21 I.A.2d *Tropical or subtropical evergreen dry "subalpine" forest.* Physiognomically resembling the winter-rain evergreen sclerophyllous dry forest (I.A.8a), usually occurring above the cloud forest (I.A.1e). Mostly evergreen sclerophyllous trees, smaller than 20 m with little or no undergrowth. (If not opened by human activity.) Poor in lianas and epiphytes, except lichens.

I.A.3 TROPICAL AND SUBTROPICAL SEMI-DECIDUOUS FOREST. Most of the upper canopy trees drought-deciduous; many of the understorey trees and shrubs evergreen and more or less sclerophyllous. However, evergreen and deciduous woody plants are not always separated by layers. They may occur mixed within the same layer, or shrubs may be primarily deciduous and trees evergreen. Nearly all trees with bud protection; leaves without "drip tips". Trees show rough bark, except some bottle trees, which may be present.

22 I.A.3a *Tropical or subtropical semi-deciduous lowland forest.* The taller trees may be bottle trees (e.g. *Ceiba*). Practically no epiphytes present. Undergrowth composed of tree reproduction and true shrubs. Succulents may be present (e.g. in form of thin-stemmed caespitose cacti). Both therophytic and hemicryptophytic lianas occur occasionally. A sparse herb layer may be present, mainly consisting of graminoid hemicryptophytes and forbs.

23 I.A.3b *Tropical or subtropical semi-deciduous montane or cloud forest.* Similar to a, the semi-deciduous lowland forest, but canopy lower and covered with xerophytic epiphytes (e.g. *Tillandsia usneoides*). Within the semi-deciduous group I.A.3, a submontane formation cannot be clearly distinguished.

I.A.4 SUBTROPICAL OMBROPHILOUS FOREST. Present only locally and in small fragmentary stands, because the subtropical climate is typically a climate with a dry season. Where the subtropical ombrophilous forest occurs, e.g. Queensland (Australia) and Taiwan, it usually grades rather inconspicuously into the tropical ombrophilous forest. Some shrubs may grow in the understorey. The subtropical ombrophilous forest should however not be confused with the tropical ombrophilous montane forest, which occurs in a climate with a similar mean annual temperature, but with less pronounced temperature differences between summer and winter. Consequently, seasonal rhythms are more evident in all subtropical forests, even in the ombrophilous ones.

The subtropical ombrophilous forest is physiognomically more closely related to the tropical than to the temperate one. Therefore the subdivisions conform more or less to the tropical ombrophilous forest, item I.A.1a to h.

24 I.A.4a
25 I.A.4b
26 I.A.4c(1)
27 I.A.4c(2)
28 I.A.4c(3)

29 I.A.4c(4)
30 I.A.4d
31 I.A.4e(1)
32 I.A.4e(2)
33 I.A.4e(3)
34 I.A.4f (1)
35 I.A.4f (2)
36 I.A.4f (3)
37 I.A.4g(1)
38 I.A.4g(2)
39 I.A.4h

40 I.A.5 MANGROVE FOREST. (Occurs only in the tidal range of the tropical and subtropical zones.) Composed almost entirely of evergreen sclerophyllous broad-leaved trees and/or shrubs with either stilt roots or pneumatophores. Epiphytes in general rare, except lichens on the branches and adnate algae on the lower parts of the trees. (Subdivisions possible; transitions to 1g, the tropical ombrophilous swamp forest, exist). E.g. coasts of Borneo, New Guinea, etc.

I.A.6 TEMPERATE AND SUBPOLAR EVERGREEN OMBROPHILOUS FOREST. Occurs only in extremely oceanic, nearly frostfree climates of the southern hemisphere, mainly in Chile. Consisting mostly of truly evergreen hemisclerophyllous trees and shrubs. Rich in thalloepiphytes and in ground-rooted herbaceous ferns.

I.A.6a *Temperate evergreen ombrophilous forest.* Vascular epiphytes and lianas may be present; height generally exceeds 10 m.

41 I.A.6a(1) Mainly broad-leaved trees, e.g. *Nothofagus* forests of New Zealand.

42 I.A.6a(2) With needle-leaved trees admixed.

43 I.A.6a(3) Mainly needle-leaved or scale-leaved trees, e.g. *Podocarpus* forests of New Zealand.

44 I.A.6b *Temperate evergreen ombrophilous alluvial forest.* Rich in forbs, e.g. western New Zealand.

I.A.6c *Temperate evergreen ombrophilous swamp forest.*

45 I.A.6c(1) Needle- or scale-leaved. Dense, tall (up to 50 m and more) scale-leaved lowland forest. Buttresses common. Closed to open ground cover of graminoids (mainly sedges) and forbs (mostly ferns). Rich in vascular and bryoid epiphytes; some lianas, e.g. *Podocarpus dacrydioides* communities of New Zealand.

46 I.A.6c(2) Broad-leaved. Tall (up to 50 m and more) broad-leaved lowland forest. Trees densely spaced so that crowns touch, but canopy permits much light to pass. Fairly open shrub synusia. Epiphytes lacking in canopy, e.g. *Eucalyptus ovata* forests of Victoria.

47 I.A.6d *Subpolar evergreen ombrophilous forest.* In contrast with I.A.6a, the temperate broad-leaved forest, vascular epiphytes lacking and canopy height much reduced (in general less than 10 m). Also leaf-size generally reduced, e.g. beech forests of New Zealand.

I.A.7 TEMPERATE EVERGREEN SEASONAL BROAD-LEAVED FOREST (with adequate summer rainfall). Consisting mainly of hemi-sclerophyllous evergreen trees and shrubs. Rich in herbaceous chamaephytic and hemicryptophytic undergrowth. Very few or no vascular epiphytes and lianas. Grades into subtropical (I.A.4) or temperate ombrophilous forests (I.A.6) or into winter-rain evergreen broad-leaved sclerophyllous forests (I.A.8). Probably includes subpolar types.

(Subdivisions possible as under tropical and subtropical seasonal forests, I.A.2a to d.)

48 I.A.7a
49 I.A.7b(1)
50 I.A.7b(2)

51	I.A.7c	
52	I.A.7d	
	I.A.8	WINTER-RAIN EVERGREEN BROAD-LEAVED SCLEROPHYLLOUS FOREST. (Often understood as Mediterranean, but present also in south-western Australia, Chile, etc. Climate with pronounced summer drought.) Consisting mainly of sclerophyllous evergreen trees and shrubs, most of which show rough bark. Herbaceous undergrowth almost lacking. No vascular and only few cryptogamic epiphytes, but evergreen woody lianas present.
53	I.A.8a	*Winter-rain evergreen sclerophyllous lowland and submontane forest.* Giant eucalypts, e.g. *Eucalyptus regnans* in Victoria and *E. diversicolor* in Western Australia. Dominated by trees over 50 m tall.
54	I.A.8b	Largely as described in I.A.8 but less than 50 m tall, e.g. Californian live-oak forests.
55	I.A.8c	Alluvial and swamp forest of this type perhaps existing, but not sufficiently known.
	I.A.9	TROPICAL AND SUBTROPICAL EVERGREEN NEEDLE-LEAVED FOREST. Consisting mainly of needle-leaved or scale-leaved evergreen trees. Broad-leaved trees may be present. Vascular epiphytes and lianas are practically lacking.
56	I.A.9a	*Tropical and subtropical lowland and submontane evergreen needle-leaved forest*, e.g. the pine forests of Honduras and Nicaragua.
57	I.A.9b	*Tropical and subtropical montane and subalpine evergreen needle-leaved forest*, e.g. the pine forests of the Philippines and southern Mexico.
	I.A.10	TEMPERATE AND SUBPOLAR EVERGREEN NEEDLE-LEAVED FOREST. Consisting mainly of needle-leaved or scale-leaved evergreen trees, but broad-leaved trees may be admixed. Vascular epiphytes and lianas practically lacking.
58	I.A.10a	*Evergreen giant forest.* Dominated by trees higher than 50 m, e.g. *Sequoia* and *Pseudotsuga* forest in the Pacific West of North America.
	I.A.10b	*Evergreen forest with rounded crowns.* Dominated by trees 45-50 m high, with more or less broad, irregularly rounded crowns, e.g. *Pinus* spp.
59	I.A.10b(1)	With evergreen sclerophyllous understorey.
60	I.A.10b(2)	Without evergreen sclerophyllous understorey.
61	I.A.10c	*Evergreen needle-leaved forest with conical crowns.* Dominated by trees 45-50 m high (only exceptionally higher), with more or less conical crowns (like most *Picea* and *Abies*), e.g. California red fir forests.
62	I.A.10d	*Evergreen forest with cylindrical crowns* (boreal). Similar to I.A.1c but crowns with very short branches and therefore very narrow, cylindro-conical.
	I.B	*MAINLY DECIDUOUS FOREST.* Majority of trees shed their foliage simultaneously in connexion with the unfavourable season.
	I.B.1	TROPICAL AND SUBTROPICAL DROUGHT-DECIDUOUS FOREST. Unfavourable season mainly characterized by drought, in most cases winter-drought. Foliage is shed regularly every year. Most trees with relatively thick, fissured bark.
63	I.B.1a	*Drought-deciduous broad-leaved lowland and submontane forest.* Practically no evergreen plants in any stratum, except some succulents. Woody and herbaceous lianas present occasionally, also deciduous bottle-trees. Ground vegetation mainly herbaceous (hemicryptophytes, particularly grasses, geophytes and some therophytes), but sparse, e.g. the broad-leaved deciduous forests of north-western Costa Rica.

64	I.B.1b	*Drought-deciduous montane (and cloud) forest.* Some evergreen species in the understorey. Drought-resistant epiphytes present or abundant, often of the bearded form (e.g. *Usnea* or *Tillandsia usneoides*). This formation is not frequent, but well developed, e.g. in northern Peru.
	I.B.2	COLD-DECIDUOUS FORESTS WITH EVERGREEN TREES (OR SHRUBS) ADMIXED. Unfavourable season mainly characterized by winter frost. Deciduous broad-leaved trees dominant, but evergreen species present as part of the main canopy or as understorey. Climbers and vascular epiphytes scarce or absent.
65	I.B.2a	*Cold-deciduous forest with evergreen broad-leaved trees and climbers* (e.g. *Ilex aquifolium* and *Hedera helix* in western Europe and Magnolia spp. in North America). Rich in cryptogamic epiphytes, including mosses. Even vascular epiphytes may be present at the base of tree stems. Lianas may be common on flood plains.
66	I.B.2b	*Cold-deciduous broad-leaved forest with evergreen needle-leaved trees*, e.g. the maple-hemlock forests of New York.
	I.B.3	COLD-DECIDUOUS FORESTS WITHOUT EVERGREEN TREES. Deciduous trees absolutely dominant. Evergreen chamaephytes and some evergreen nanophanerophytes may be present. Climbers insignificant but may be common on flood plains, vascular epiphytes absent (except occasionally at the lower base of the tree); thalloepiphytes always present, particularly lichens.
67	I.B.3a	*Temperate lowland and submontane broad-leaved cold-deciduous forest.* Trees up to 50 m tall, e.g. the Mixed mesophytic Forest (U.S.A.). Primarily algae and crustose lichens as epiphytes.
	I.B.3b	*Montane or boreal cold-deciduous forest* (including lowland or submontane in topographic positions with high atmospheric humidity). Foliose and fruticose lichens, and bryophytes as epiphytes. Trees up to 50 m tall, but in montane or boreal forest normally not taller than 30 m.
68	I.B.3b(1)	Mainly broad-leaved deciduous.
69	I.B.3b(2)	Mainly needle-leaved deciduous (e.g. Siberian *Larix*) forests.
70	I.B.3b(3)	Mixed broad-leaved and needle-leaved deciduous.
	I.B.3c	*Subalpine or subpolar cold-deciduous forest.* In contrast to I.B.3a and b, the lowland and montane cold-deciduous forest, the canopy height significantly reduced (not taller than 20 m). Tree trunks frequently gnarled. Epiphytes similar as in b, but in general more abundant. Often grading into woodland (see Formation Class II).
71	I.B.3c(1)	With primarily hemicryptophytic undergrowth.
72	I.B.3c(2)	With primarily chamaephytic undergrowth. May merge with forests admixed with conifers.
	I.B.3d	*Cold-deciduous alluvial forest.* Flooded by rivers, therefore moister and richer in nutrients than cold-deciduous lowland forest, I.B.3a. Trees and shrubs with high growth rates and vigorous herbaceous undergrowth.
73	I.B.3d(1)	Occasionally flooded; physiognomically similar to I.B.3a, with tall trees and abundant macrophyllous shrubby undergrowth.
74	I.B.3d(2)	Regularly flooded; trees not as tall and dense as in I.B.3a, but herbaceous undergrowth abundant and tall. (In Eurasia *Salix*- or *Alnus*-species frequently dominating.)
	I.B.3e	*Cold-deciduous swamp or peat forest.* Flooded until late spring or early summer, surface soil organic. Relatively poor in tree species. Ground cover of varied growth forms mostly continuous. (Subdivisions as under I.B.3b, the boreal cold-deciduous forest).

75 I.B.3e(1)
76 I.B.3e(2)
77 I.B.3e(3)

I.C *EXTREMELY XEROMORPHIC FOREST.* Dense stands of xeromorphic phanerophytes, such as bottle trees, tuft trees with succulent leaves and stem succulents. Undergrowth with shrubs of similar xeromorphic adaptations, succulent chamaephytes and herbaceous hemicryptophytes, geophytes and therophytes. Often grading into woodlands. (See Formation Class II.)

78 I.C.1 SCLEROPHYLLOUS-DOMINATED EXTREMELY XEROMORPHIC FOREST. Life form combination as above, except for predominance of sclerophyllous trees, many of which have bulbose stem bases largely embedded in the soil (xylopods).

I.C.2 THORN-FOREST. Species with thorny appendices predominate.

79 I.C.2a *Mixed deciduous-evergreen thorn forest.*

80 I.C.2b *Purely deciduous thorn forest.*

81 I.C.3 MAINLY SUCCULENT FOREST. Tree-formed (scapose) and shrub-formed (caespitose) succulents very frequent, but other xero-phanerophytes are usually present as well.

II WOODLAND (open stands of trees)

Composed of trees at least 5 m tall with crowns not usually touching but with a coverage of at least 40 per cent. A herbaceous synusia may be present. See Formation Group V.A.1 if the coverage of the trees is less than 40 per cent and there is a herbaceous synusia. The boundary of 40 per cent coverage is convenient because it can be estimated with ease during the field-work: when the coverage of the trees is 40 per cent the distance between two tree crowns equals the mean radius of a tree crown.

II.A *MAINLY EVERGREEN WOODLAND,* i.e. evergreen as defined in I.A.

82 II.A.1 EVERGREEN BROAD-LEAVED WOODLAND. Mainly sclerophyllous trees and shrubs, no epiphytes.

II.A.2 EVERGREEN NEEDLE-LEAVED WOODLAND. Mainly needle- or scale-leaved. Crowns of many trees extending to the base of the stem or at least very branchy.

II.A.2a *Evergreen needle-leaved woodland with rounded crowns* (e.g. *Pinus*).
83 II.A.2a(1) With evergreen sclerophyllous understorey (Mediterranean).
84 II.A.2a(2) Without evergreen sclerophyllous understorey.

85 II.A.2b *Evergreen needle-leaved woodland with conical crowns prevailing* (mostly subalpine).

86 II.A.2c *Evergreen needle-leaved woodland with very narrow cylindro-conical crowns* (e.g. *Picea* in the boreal region).

II.B *MAINLY DECIDUOUS WOODLAND* (see I-B).

II.B.1 DROUGHT-DECIDUOUS WOODLAND.
(Subdivisions more or less like forests.)

87 II.B.1a
88 II.B.1b

	II.B.2	COLD-DECIDUOUS WOODLAND WITH EVERGREEN TREES (see I.B.2).
89	II.B.2a	
90	II.B.2b	
	II.B.3	COLD-DECIDUOUS WOODLAND WITHOUT EVERGREEN TREES (see I.B.3). Most frequent in the subarctic region, elsewhere only on swamps or bogs.
91	II.B.3a	*Broad-leaved deciduous woodland.*
92	II.B.3b	*Needle-leaved deciduous woodland.*
93	II.B.3c	*Mixed deciduous woodland* (broad-leaved and needle-leaved).
	II.C	*EXTREMELY XEROMORPHIC WOODLAND.* Similar to I.C, the only difference being the more sparse stocking of individual trees. (Subdivisions as under I.C.)
94	II.C.1	
95	II.C.2a	
96	II.C.2b	
97	II.C.3	

	III	SCRUB (shrubland or thicket)
		Mainly composed of caespitose woody phanerophytes 0.5 to 5 m tall.[1] Each of the following subdivisions may be either: Shrubland: most of the individual shrubs not touching each other; often with a grass stratum; or Thicket: individual shrubs interlocked.
	III.A	*MAINLY EVERGREEN SCRUB* (evergreen in the sense of I.A).
	III.A.1	EVERGREEN BROAD-LEAVED SHRUBLAND (or thicket).
98	III.A.1a	*Low bamboo thicket* (or, less frequently, shrubland). Lignified creeping graminoid nano- or microphanerophytes.
99	III.A.1b	*Evergreen tuft-tree shrubland* (or thicket). Composed of small trees and woody shrubs (e.g. Mediterranean dwarf palm shrubland or Hawaiian tree fern thicket).
100	III.A.1c	*Evergreen broad-leaved hemisclerophyllous thicket* (or shrubland). Caespitose, creeping or lodged nano- or microphanerophytes with relatively large and soft leaves (e.g. subalpine *Rhododendron* thickets, or *Hibiscus tiliaeceus* matted thickets of Hawaii).
101	III.A.1d	*Evergreen broad-leaved sclerophyllous shrubland* (or thicket). Dominated by broad-leaved sclerophyllous shrubs and immature trees (i.e. chaparral or macchia). May often merge with parkland, grassland or heath.
102	III.A.1e	*Evergreen suffruticose thicket* (or shrubland). Stand of semi-lignified nanophanerophytes that in dry years may shed part of their shoot systems (e.g. *Cistus* heath).[2]
	III.A.2	EVERGREEN NEEDLE-LEAVED AND MICROPHYLLOUS SHRUBLAND (or thicket).
103	III.A.2a	*Evergreen needle-leaved thicket* (or shrubland). Composed mostly of creeping or lodged needle-leaved phanerophytes (e.g. *Pinus mughus*, "Krummholz").

1. Not to be confused with developing second growth forests, see footnote relating to I. Sometimes, scrub may reach more than 5 m in height.
2. Occasionally less than 50 cm tall, thereby grading into IV.A.1a.

104	III.A.2b	*Evergreen microphyllous shrubland* (or thicket). Often ericoid shrubs (mostly in tropical subalpine belts).
	III.B	*MAINLY DECIDUOUS SCRUB* (deciduous in the sense of I.B).
105	III.B.1	DROUGHT-DECIDUOUS SCRUB WITH EVERGREEN WOODY PLANTS ADMIXED.
106	III.B.2	DROUGHT-DECIDUOUS SCRUB WITHOUT EVERGREEN WOODY PLANTS ADMIXED.
	III.B.3	COLD-DECIDUOUS SCRUB.
107	III.B.3a	*Temperate deciduous thicket* (or shrubland). More or less dense scrub without or with only little herbaceous undergrowth. Poor in cryptogams.
	III.B.3b	*Subalpine or subpolar deciduous thicket* (or shrubland). Upright or lodged caespitose nanophanerophytes with great vegetative regeneration capacity. As a rule completely covered by snow for at least half a year.
108	III.B.3b(1)	With primarily hemicryptophytic undergrowth, mainly forbs (e.g. subalpine *Alnus viridis* thicket).
109	III.B.3b(2)	With primarily chamaephytic undergrowth, mainly dwarf shrubs and fruticose lichens (e.g. *Betula tortuosa* shrubland at the polar tree line).
	III.B.3c	*Deciduous alluvial shrubland* (or thicket). Fast-growing shrubs, occurring as pioneers on river banks or islands that are often vigorously flooded, therefore mostly with very sparse undergrowth.
110	III.B.3c(1)	With lanceolate leaves (e.g. *Salix*, mostly in lowland or submontane region).
111	III.B.3c(2)	Microphyllous.
112	III.B.3d	*Deciduous peat shrubland* (or thicket). Upright caespitose nanophanerophytes with *Sphagnum* and (or) other peat mosses.
	III.C	*EXTREMELY XEROMORPHIC (SUBDESERT) SHRUBLAND.* Very open stands of shrubs with various xerophytic adaptations, such as extremely scleromorphic or strongly reduced leaves, green branches without leaves, or succulent stems, etc., some of them with thorns.
	III.C.1	MAINLY EVERGREEN SUBDESERT SHRUBLAND. In extremely dry years some leaves and shoot portions may be shed.
	III.C.1a	*Evergreen subdesert shrubland.*
113	III.C.1a(1)	Broad-leaved, dominated by sclerophyllous nanophanerophytes, including some phyllocladous shrubs (e.g. mulga scrub in Australia).
114	III.C.1a(2)	Microphyllous, or leafless, but with green stems (e.g. *Retama retam*).
115	III.C.1a(3)	Succulent, dominated by variously branched stem and leaf succulents.
	III.C.1b	*Semi-deciduous subdesert shrubland.* Either facultatively deciduous shrubs or a combination of evergreen and deciduous shrubs.
116	III.C.1b(1)	Facultatively deciduous (e.g. *Atriplex-Kochia*-saltbush in Australia and North America).
117	III.C.1b(2)	Mixed evergreen and deciduous, transitional to III.C.2.
	III.C.2	DECIDUOUS SUBDESERT SHRUBLAND. Mainly deciduous shrubs, often with a few evergreens.
118	III.C.2a	*Deciduous subdesert shrubland without succulents.*
119	III.C.2b	*Deciduous subdesert shrubland with succulents.*

IV DWARF-SCRUB AND RELATED COMMUNITIES

Rarely exceeding 50 cm in height (sometimes called heaths or heathlike formations). According to the density of the dwarf-shrub cover are distinguished:
Dwarf-shrub thicket: branches interlocked;
Dwarf-shrubland: individual dwarf-shrubs more or less isolated or in clumps;
Cryptogamic formations with dwarf-shrubs: surface densely covered with mosses or lichens (thallochamaephytes); dwarf-shrubs occurring in small clumps or individually. In the case of bogs locally dominating graminoid communities may be included.

IV.A *MAINLY EVERGREEN DWARF-SCRUB.* Most dwarf-shrubs evergreen.

IV.A.1 EVERGREEN DWARF-SHRUB THICKET. Densely closed dwarf-shrub cover, dominating the landscape ("dwarf-shrub heath" in the proper sense).

120 IV.A.1a *Evergreen caespitose dwarf-shrub thicket.* Most of the branches standing in upright position, often occupied by foliose lichens. On the ground pulvinate mosses, fruticose lichens or herbaceous life forms may play a role (e.g. *Calluna* heath).

121 IV.A.1b *Evergreen creeping or matted dwarf-shrub thicket.* Most branches creeping along the ground. Variously combined with thallochamaephytes in which the branches may be embedded (e.g. *Loiseleuria* heath).

IV.A.2 EVERGREEN DWARF-SHRUBLAND. Open or more loose cover of dwarf-shrubs.

122 IV.A.2a *Evergreen cushion shrubland.* More or less isolated clumps of dwarf-shrubs forming dense cushions, often equipped with thorns (e.g. *Astragalus-* and *Acantholimon* "porcupine"-heath of the East Mediterranean mountains).

IV.A.3 MIXED EVERGREEN DWARF-SHRUB AND HERBACEOUS FORMATION. More or less open stands of evergreen suffrutescent or herbaceous chamaephytes, various hemicryptophytes, geophytes, etc.

123 IV.A.3a *Truly evergreen dwarf-shrub and herb mixed formation* (e.g. *Nardus-Calluna*-heath).

124 IV.A.3b *Partially evergreen dwarf-shrub and herb mixed formation.* Many individuals shed parts of their shoot systems during the dry season (e.g. *Phrygana* in Greece).

IV.B *MAINLY DECIDUOUS DWARF-SCRUB.* Similar to IV.A, but mostly consisting of deciduous species.

125 IV.B.1 FACULTATIVELY DROUGHT-DECIDUOUS DWARF-THICKET (or dwarf-shrubland). Foliage is shed only in extreme years.

IV.B.2 OBLIGATORY, DROUGHT-DECIDUOUS DWARF-THICKET (or dwarf-shrubland). Densely closed dwarf-shrub stands which lose all or at least part of their leaves in the dry season.

126 IV.B.2a *Drought-deciduous caespitose dwarf-thicket.* Corresponding to IV.A.1a.

127 IV.B.2b *Drought-deciduous creeping or matted dwarf-thicket.* Corresponding to IV.A.1b.

128 IV.B.2c *Drought-deciduous cushion dwarf-shrubland.* Corresponding to IV.A.2a.

129 IV.B.2d *Drought-deciduous mixed dwarf-shrubland.* Deciduous and evergreen dwarf-shrubs, caespitose hemicryptophytes, succulent chamaephytes and other life forms intermixed in various patterns.

IV.B.3 COLD-DECIDUOUS DWARF-THICKET (or dwarf-shrubland). Physiognomically similar to IV.B.2, but shedding the leaves at the beginning of a cold season. Usually richer in cryptogamic chamaephytes.

(Subdivisions similar to IV.B.2. Transitions into IV.D and E possible.)

130 IV.B.3a
131 IV.B.3b
132 IV.B.2c
133 IV.B.2d

IV.C *EXTREMELY XEROMORPHIC DWARF-SHRUBLAND.* More or less open formations consisting of dwarf-shrubs, succulents, geophytes, therophytes and other life forms adapted to survive or to avoid a long dry season. Mostly subdesertic.

(Subdivisions similar to III-C, extremely xeromorphic shrublands.)

134 IV.C.1a(1)
135 IV.C.1a(2)
136 IV.C.1a(3)
137 IV.C.1b(1)
138 IC.V.1b(2)
139 IV.C.2a

IV.D *TUNDRA.* Slowly growing, low formations, consisting mainly of dwarf-shrubs, graminoids, and cryptogams, beyond the subpolar tree line. Often showing plant patterns caused by freezing movements of the soil (cryoturbation). Except in boreal regions, dwarf-shrub formations above the mountain tree line should not be called tundra, because they are as a rule richer in dwarf-shrubs and grasses, and grow taller due to the greater radiation in lower latitudes.

IV.D.1 MAINLY BRYOPHYTE TUNDRA. Dominated by mats or small cushions of chamaephytic mosses. Groups of dwarf-shrubs are as a rule scattered irregularly and are not very dense. General aspect more or less dark green, olive green or brownish.

141 IV.D.1a *Caespitose dwarf-shrub—moss tundra.*

142 IV.D.1b *Creeping or matted dwarf-shrub—moss tundra.*

143 IV.D.2 MAINLY LICHEN TUNDRA. Mats of fruticose lichens dominating, giving the formation a more or less pronounced grey aspect. Dwarf-shrubs mostly evergreen, creeping or pulvinate. Dwarf-shrub—lichen tundra.

IV.E *MOSSY BOG FORMATIONS WITH DWARF-SHRUB.* Oligotrophic peat accumulations formed mainly by *Sphagnum* or other mosses, which as a rule cover the surface as well. Dwarf-shrubs are concentrated on the relatively drier parts or are loosely scattered. To a certain extent they resemble dwarf-shrub formations on mineral soil. Graminoid hemicryptophytes, geophytes with rhizomes and other herbaceous life forms may dominate locally. Slowly growing trees and shrubs can grow as isolated individuals, in groups or in woodlands, which are marginal to the bog or may be replaced by open formations in a cyclic succession. The following subdivisions correspond to the classification of bog types adopted in Europe.

IV.E.1 RAISED BOG. By growth of *Sphagnum* species raised above the general ground-water table and having a ground-water table of their own. Therefore not supplied by "mineral" water (i.e. water having been in touch with the inorganic soil), but only by rain water (truly ombrotrophic bogs).

144 IV.E.1a *Typical raised bog* (suboceanic, lowland and submontane). Mosses dominating throughout, except on locally raised dry hummocks, which are dominated by dwarf-shrubs. Trees rare and, if present, concentrated on the marginal slopes of the convex peat accumulation. Mostly surrounded by a very wet, but less oligotrophic swamp.

145 IV.E.1b *Montane (or "subalpine") raised bog.* Growing slower than the typical raised bog (or formed in an earlier period with a warmer climate and actually "dead" or being destroyed by erosion). Often covered with sedges or evergreen dwarf-shrubs. Micro- or nanophanerophytes (e.g. *Pinus mughus*) locally dominating.

146 IV.E.1c *Subcontinental woodland bog.* Temporarily covered by open wood of low productivity, which in a sequence of wetter years may be replaced by *Sphagnum* formations similar to IV.E.1a.

IV.E.2 NON-RAISED BOG. Not or not very markedly raised above the mineral-water table of the surrounding landscape. Therefore in general wetter and not as oligotrophic as IV.E.1. Poorer in mosses than the typical raised bog, 1a, to which various forms of transitions are possible.

147 IV.E.2a *Blanket bog* (oceanic lowland, submontane or montane). The micro-surface of the bog is less undulating and less rich in actively growing mosses than in IV.E.1a. Evergreen dwarf-shrubs are scattered as well as caespitose hemicryptophytes (sedges or grasses) and some rhizomatous geophytes.

148 IV.E.2b *String bog* (Finnish "aapa" bog). Flat oligotrophic with string-like hummocks in the boreal lowlands. The Finnish name indicates an open bog without or with only a few trees of very poor vigour, which grow on narrow and low elongated hummocks, the so-called strings. These peat strings are formed by pressure of the ice covering the more or less flooded bog from early fall to late spring. Only these strings are covered by dwarf-shrubs and are rich in *Sphagnum*. The main part of the bog is similar to a wet sedge swamp.

Herbaceous vegetation

The classification of herbaceous vegetation requires special consideration mainly because of (a) the continual seasonal changes in the physiognomy of the communities; (b) the problems of distinguishing many tropical from non-tropical grasslands; (c) the management of grasslands which can seriously affect the structure of the vegetation and which can be changed frequently; (d) the problems of distinguishing between natural and managed grasslands.

There are two major types of herbaceous growth forms: graminoids and forbs. Graminoids include all herbaceous grasses and grasslike plants such as sedges (*Carex*), rushes (*Juncus*), cat's-tails (*Typha*), etc. Forbs are broad-leaved herbaceous plants such as clover (*Trifolium*), sunflowers (*Helianthus*) ferns, milkweeds (*Asclepias*), etc. Usually all non-graminoid herbaceous plants are included in the forbs.

The Unesco classification frequently employs the term grassland to signify a herbaceous type of vegetation dominated by graminoid growth forms. As the graminoids include numerous taxa other than grasses (*Gramineae*), the term grassland must here be understood to mean a physiognomic vegetation type without floristic implications.

As with woody communities, height is a major feature to characterize a plant community dominated by herbaceous growth forms. The great seasonal fluctuations in the height of herbaceous plants require that height measurements (or estimates) be made at the time of flowering, i.e. when the inflorescences are developed. Inflorescenses may not develop where steady or heavy grazing prevails; their height must then be estimated.

Coverage is another important factor in characterizing herbaceous vegetation. However, at the scale of 1 : 1,000,000 or less, all herbaceous communities are assumed to have a more or less continuous coverage, and no mention is made of this in the map legend. The exception is very low density. In that case, the community is called open.

The growth forms of graminoids are significant and it is common to distinguish sod forms, bunch (tussock) forms and combinations of these. In the Unesco classification, all graminoid communities are considered to be more or less of the sod form, and this need therefore not be mentioned. However, when bunch grasses dominate, they affect the physiognomy of the vegetation profoundly. The bunch grass form must then be included in the description of the plant communities involved. Such descriptions parallel the descriptions of the crown forms of needle-leaved evergreen trees, as for instance in items I.A.9c and d.

Herbaceous vegetation types often include a synusia of woody plants that gives the type a special character. Frequently, therefore, such synusiae are used to characterize the herbaceous communities. Some of the more important features of a woody synusia include height and density, whether evergreen or deciduous, needle-leaved, broad-leaved or aphyllous (essentially without leaves), etc. Such features prevail in a great variety of combinations and graphically describe the general physiognomy of the vegetation. Other features are admissible if they characterize a community that is extensive enough to be mapped at the scale of 1 : 1,000,000 or less, e.g. the sclerophyllous nature of the woody synusia of many *Eucalyptus*-grass combinations in Australia.

There are numerous, frequently used vegetation terms such as savannah, steppe, meadow, etc. They have been avoided in the definitions of the Unesco classification because they have too many conflicting interpretations. Occasionally, however, they have been added in parentheses where this helps the reader to identify the category. It is always best to use the Unesco terminology on the vegetation maps. Locally established terms meaningful to the inhabitants of the respective regions (e.g. *campo cerrado*) may be added to but do not replace the Unesco terms in the map legends. In this manner, vegetation maps are meaningful to local users as well as to a world-wide audience. This is especially important for comparative studies.

V HERBACEOUS VEGETATION

V.A *TALL GRAMINOID VEGETATION.* In tall grasslands, the dominant graminoid growth forms are over 2 m tall when the inflorescences are fully developed. Forbs may be present but their coverage is less than 50 per cent.

V.A.1 TALL GRASSLAND WITH A TREE SYNUSIA COVERING 10-40 PER CENT, with or without shrubs. This is somewhat like a very open woodland with a more or less continuous ground cover (over 50 per cent) of tall graminoids. For categories with a tree synusia covering over 40 per cent see Formation Class II.

149 V.A.1a *Woody synusia broad-leaved evergreen.*

150 V.A.1b *Woody synusia broad-leaved semi-evergreen*, i.e. composed of at least 25 per cent each of broad-leaved evergreen and broad-leaved deciduous trees.

V.A.1c *Woody synusia broad-leaved deciduous.*

151 V.A.1c(1) Similar to the above but seasonally flooded, e.g. in north-east Bolivia.

V.A.2 TALL GRASSLAND WITH A TREE SYNUSIA COVERING LESS THAN 10 PER CENT, with or without shrubs. Subdivisions V.A.2a to c as in V.A.1.

152 V.A.2a

153 V.A.2b

154 V.A.2c

155 V.A.2d *Tropical or subtropical tall grassland with trees and/or shrubs growing in tufts on termite nests* (termite savannah).

	V.A.3	TALL GRASSLAND WITH A SYNUSIA OF SHRUBS (shrub savannah). Subdivisions as in V.A.2.
156	V.A.3a	
157	V.A.3b	
158	V.A.3c	
159	V.A.3d	
	V.A.4	TALL GRASSLAND WITH A WOODY SYNUSIA CONSISTING MAINLY OF TUFT PLANTS (usually palms).
160	V.A.4a	*Tropical grassland with palms*, e.g. the palm savannas of *Arocomia totai* and *Attalea princeps* north of Santa Cruz de la Sierra, Bolivia.
161	V.A.4a(1)	Similar to above, seasonally flooded, e.g. the *Mauritia vinifera* savannahs, Llanos de Mojos, Bolivia.
	V.A.5	TALL GRASSLAND PRACTICALLY WITHOUT WOODY SYNUSIA.
162	V.A.5a	*Tropical grassland* as described in V.A.5 as in various low-latitude regions of Africa.
163	V.A.5a(1)	Similar to above, seasonally flooded, e.g. the Campos de Várzea of the lower Amazon Valley.
164	V.A.5a(2)	Similar to above, wet or flooded most of the year, e.g. the papyrus (*Cyperus papyrus*) swamps of the upper Nile Valley.
	V.B	*MEDIUM TALL GRASSLAND*, i.e. the dominant graminoid growth forms are 50 cm to 2 m tall when their inflorescences are fully developed. Forbs may be present but cover less than 50 per cent. Divided and subdivided as V.A.1 to V.A.3d.
165	V.B.1a	
166	V.B.1b	
167	V.B.1c	
168	V.B.2a	
169	V.B.2b	
170	V.B.2c	
171	V.B.2d	
172	V.B.3a	
173	V.B.3b	
174	V.B.3c	
175	V.B.3d	
176	V.B.3e	*Woody synusia consisting mainly of deciduous thorny shrubs*, e.g. the tropical thorn bush savannah of the Sahel region in Africa with *Acacia tortilis*, *A. senegal* and others.
	V.B.4	MEDIUM TALL GRASSLAND WITH AN OPEN SYNUSIA OF TUFT PLANTS, usually palms.
177	V.B.4a	*Medium tall subtropical grassland with open groves of palms*, e.g. in Corrientes, Argentina.
178	V.B.4a(1)	Similar to the above, seasonally flooded, e.g. *Mauritia* palm groves in the Colombian and Venezuelan llanos.
	V.B.5	MEDIUM TALL GRASSLAND PRACTICALLY WITHOUT WOODY SYNUSIA.
179	V.B.5a	*Medium tall grassland consisting mainly of sod grasses*, e.g. the tall-grass prairie in eastern Kansas.
180	V.B.5a(1)	Wet or flooded most of the year, e.g. *Typha* swamps.
181	V.B.5a(2)	On sandy soil or dunes, e.g. communities of *Andropogon hallii* in the Nebraska Sand Hills.

182 V.B.5b *Medium tall grassland consisting mainly of bunch grasses*, e.g. the hard tussock (*Festuca novae-zelandiae*) grasslands in New Zealand.

V.C *SHORT GRASSLAND*, i.e. the dominant graminoid growth forms are less than 50 cm tall when their inflorescences are developed. Forbs may be present but they cover less than 50 per cent. Divided and subdivided as V.B.1 to V.B.4.

183 V.C.1a
184 V.C.1b
185 V.C.1c

186 V.C.2a
187 V.C.2b
188 V.C.2c
189 V.C.2d

190 V.C.3a
191 V.C.3b
192 V.C.3c
193 V.C.3d
194 V.C.3e

195 V.C.4a
196 V.C.4a(1)

197 V.C.5a *Tropical alpine open to closed bunch-grass communities with a woody synusia of tuft plants (Espeletia, Lobelia, Senecio)*, microphyllous to leptophyllous dwarf-shrubs and cushion plants, often with woolly leaves. Above timberline in low latitudes: Páramo and related vegetation types without snow in the alpine regions of Kenya, Colombia, Venezuela, etc.

198 V.C.5b Similar to V.C.5a but very open and without tuft plants. Frequent nocturnal snowfall (snow gone by 9 a.m.), the Super-Páramo (i.e. above Páramo) of J. Cuatrecasas.[1]

199 V.C.5c *Tropical or subtropical alpine bunch-grass vegetation with open stands of evergreen with or without deciduous shrubs and dwarf shrubs*, e.g. Puna.

V.C.5c(1) With numerous succulent growth forms.

200 V.C.5d *Bunch grass vegetation of varying coverage with dwarf shrubs.*

V.C.5d(1) With cushion plants that may be locally more important than the dwarf shrubs, e.g. Puna south of Oruro, Bolivia.

V.C.6 SHORT GRASSLANDS PRACTICALLY WITHOUT WOODY SYNUSIA.

201 V.C.6a *Short-grass communities.* They may fluctuate in structure and floristic composition due to greatly fluctuating precipitation of semi-arid climate, e.g. short-grass (*Bouteloua gracilis* and *Buchloë dactyloides*) prairie of eastern Colorado.

202 V.C.6b *Bunch-grass communities*, e.g. blue tussock (*Poa colensoi*) communities of New Zealand, and alpine dry Puna with *Festuca orthophylla* of northern Chile and southern Bolivia.

V.C.7 SHORT TO MEDIUM TALL MESOPHYTIC GRASSLAND (meadows).

203 V.C.7a *Sodgrass communities*, often rich in forbs, usually dominated by hemicryptophytes. Occurring chiefly in lower altitudes with a cool, humid climate in North America and

1. "Páramo Vegetation and its Life Forms", *Colloquium Geographicum*, Vol. 9, Bonn, F. Dümmlers Verlag, 1968, 223 p.

Eurasia. Many plants may remain at least partly green during the winter, even below the snow in the higher latitudes.

204	V.C.7b	*Alpine and subalpine meadows of the higher latitudes*, in contrast to the Páramo and Puna types of vegetation in the lower latitudes. Usually moist much of the summer due to melt water.
205	V.C.7b(1)	Rich in forbs, e.g. on the Olympic Peninsula, Washington.
206	V.C.7b(2)	Rich in dwarf shrubs, e.g. on the Rocky Mountains of Colorado.
207	V.C.7b(3)	Snow-bed communities. Open communities, rich in small forbs and/or forblike dwarf shrubs (e.g. *Salix herbacea*). High latitude equivalent of the low latitude super-Páramo (cf. V.C.5b).
208	V.C.7b(4)	Avalanche meadows, occurring as narrow strips of grassland between forests on steep slopes of high mountains where avalanches, descending annually in spring, prevent forest growth. Variable in structure; may have some shrubs and damaged trees.
	V.C.8	GRAMINOID TUNDRA. As in the case of the dwarf-shrub tundra (Formation subclass IV.D), the use of the term graminoid tundra should be limited to high latitudes, i.e. beyond the polar timberline.
209	V.C.8a	*Graminoid bunch-form tundra.* Most graminoids grow in tussock form. Mosses and/or lichens often grow between the tussocks. *Eriophorum* tundra in northern Alaska.
210	V.C.8a(1)	Seasonally flooded.
211	V.C.8b	*Graminoid sod-form tundra.* The graminoids form a more or less dense sod, often with a variety of forbs and dwarf shrubs (both caespitose and creeping). Lichens may be common. Tundra near Lake Iliamna, Alaska.
212	V.C.8b(1)	Seasonally flooded.
	V.D	*FORB VEGETATION.* The plant communities consist mainly of forbs, i.e. the coverage of forbs exceeds 50 per cent. Graminoids may be present but they cover less than 50 per cent, often much less.
	V.D.1	TALL FORB COMMUNITIES. The dominant forb growth forms are more than 1 m tall when fully developed.
213	V.D.1a	*Mainly perennial flowering forbs, and ferns.*
214	V.D.1a(1)	Saline substrate, and/or wet much of the year, e.g. the *Batis-Salicornia* marshes of Florida and the *Heliconia-Calathea* formations of Central America.
215	V.D.1b	*Fern thickets*, sometimes in nearly pure stands, especially in humid climates, e.g. *Pteridium aquilinum.*
216	V.D.1c	*Mainly annual forbs.*
	V.D.2	LOW FORB COMMUNITIES, dominated by forb growth forms less than 1 m tall when fully developed.
217	V.D.2a	*Mainly perennial flowering forbs, and ferns*, e.g. the *Celmisia* meadows in New Zealand and the Aleutian forb meadows, Alaska.
218	V.D.2b	*Mainly annual forbs.*
219	V.D.2b(1)	Ephemeral forb communities in tropical and subtropical regions with very little precipitation where, from autumn to spring, clouds moisten vegetation and soil, e.g. in the coastal hills of Peru and northern Chile. Dominated by annual forbs which germinate at the beginning of the cloudy season, growing abundantly until the end of it, and giving the landscape a fresh green look. The dry season aspect is desert-like. Geophytes and cryptogamic hemicryptophytes or chamaephytes are always present

and may dominate locally. Some phanerophytes may occur as relicts of natural cloud woodland.

220 V.D.2b(2) Ephemeral or episodical forb communities of arid regions: the "flowering desert". Mostly fast growing forbs, sometimes concentrated in depressions where water can accumulate. Sometimes these communities can form ground synusiae in shrub or dwarf shrub formations of arid regions (cf. IV.C), e.g. in Sonoran Desert.

221 V.D.2b(3) Episodical forb communities. Irregular communities of varying structure and composition, developing in the dry parts of river beds during low-water periods of more than two months, or on temporarily dry areas in beds where the river tends to change channels more or less frequently, e.g. in the river beds of the major rivers of the llanos of Colombia and Venezuela.

V.E *HYDROMORPHIC FRESH-WATER VEGETATION* (aquatic vegetation). As in the case of the aquatic mangrove communities, so helophyte communities are not classified in this category. Thus, stands of *Typha* are considered medium tall graminoid communities, wet or flooded most of the year (cf. V.B.5a(1)).

V.E.1 ROOTED FRESH-WATER COMMUNITIES. Composed of aquatic plants that are structurally supported by the water, i.e. not self-supporting in contrast to helophytes.

222 V.E.1a *Tropical and subtropical forb formations* without appreciable seasonal contrasts, e.g. *Victoria regia* communities of the Amazon.

223 V.E.1b *Middle and higher latitude forb formations* with major seasonal contrasts, e.g. *Nymphaea* and *Nuphar* communities.

V.E.2 FREE-FLOATING FRESH-WATER COMMUNITIES.

224 V.E.2a *Tropical and subtropical free-floating formations*, e.g. *Pistia*, *Eichhornia*, *Azolla pinnata*, etc.

225 V.E.2b *Free-floating formations of the middle and higher latitudes*, disappearing during the winter, e.g. *Utricularia purpurea*, *Lemna*.

Cartography

LIMITS

The various units we have distinguished between above must be delimited on the map by a line separating them from neighbouring units. On a geological map this is easy enough. But in the case of climatic or vegetation maps, this is often more difficult to indicate as the units to be represented are rarely very clearly delimited. This is true for a forest or a swamp; but often there is an uninterrupted transition from one unit to the next. Typographically it is very difficult to represent gradations and very often limits are drawn with an accuracy which does not exist in nature.

In the case of a scale of 1 : 5,000,000, the contours given by the authors of maps already published are sometimes not accurate enough. Aerial photographs, especially when taken at a high altitude, can play a useful part in fixing the contours of formations whose botanical characteristics are known.

For larger scales, aerial photographs are indispensable. Naturally, the topographical background must be accurate, which is not always achieved.

COLOURS TO BE USED

If we adopt the ideas put into practice since 1924 by H. Gaussen, the theoretical principles are as follows:

1. The way the colour is put on (flat tints, shading with lines, shading with dots) represents the basic physiognomic types: forest, shrubland, herbaceous vegetation.
2. The colour used indicates the environmental conditions through a combination of basic colours each of which represents a feature of the surroundings. This procedure, first carried out in 1926, has given good results in the Unesco-FAO 1 : 5,000,000 maps of Mediterranean countries, the 1 : 1,000,000 maps of India, the Sahara and Madagascar and the 1 : 200,000 maps of vegetation in France and North Africa.

Once the reader has become familiar with the choice of colours, an idea of the general vegetation conditions will be obtained at a glance. These conditions are basically climatic for small-scale maps (1 : 5,000,000), but the prevalence of certain types of soil may be indicated (dunes, mangroves, swamps, lakes).

A folder showing the Unesco classification, with samples of the main colours, is attached to this notice.

In tropical countries, where there is a considerable difference between environmental conditions in the dry and rainy seasons, alternating vertical strips are used representing both types.

For each colour, three different grades of intensity are used: e.g. in the case of blue "B": B1, very light; B2, fairly heavy wash; B3, flat tint.

The basic colours are: Red = R, Orange = O, Yellow = Y, Green = V (*vert* in French), Blue = B, Violet = P (purple-violet), and Grey = G, plus Brown (*marron*) = M for peat-bogs. Numerous shades are formed by the superimposition of colours.

The ecological colour indicated by the sequence number (the bold figures in the left-hand column in the Classification) is obtained by superimposing primary colours representing different factors of the environment. At the scale of 1 : 5,000,000 the main factors are humidity and heat.

The colour scale is the following, according to the colours of the rainbow.

Humidity factor
Very dry: orange
Medium: yellow
Very humid: blue
} with all the intermediate shades

Heat factor
Very hot: red
Medium: yellow
Very cold: grey
} with all the intermediate shades

A temperate forest, i.e. with little humidity and medium heat is: light blue + yellow = light green.
A tropical forest, i.e. humid and hot is: blue + red = violet.
A hot desert, i.e. dry and hot is: red + orange = orange-red.
A cold desert, i.e. dry and cold is: orange + grey = orange-grey, etc.
The lack of vegetation can be indicated as follows: white with heat: little red dots; white very cold: little grey dots.

Sometimes the nature of the soil must be taken into account: sand, bogs, mangrove, etc. This can be represented by special signs.

SYMBOLS USED

Symbols are based upon those published by H. Gaussen and accepted by Braun-Blanquet. They are used for large-scale maps. Each genus possesses a certain type of symbol and the various species are variations on this prototype.

In small-scale physiognomic maps, it is impossible to enter into so much detail.

Different types of growth are distinguished: broad-leaved trees, tall, medium-tall, bushy. Similarly in the case of conifers.

Supplementary symbols: thorny, succulent, microphyllous, peat bogs, etc.

Deciduous vegetation is represented without any blacked-in area.

Evergreen vegetation is drawn in black.

Savannahs and meadows each have their special symbols.

Swamps, fresh water, salt water (mainland) also have their own symbols.

Annex concerning the representation of cultivated areas

In cartographical and geographical matters in general, scale is of the first importance.

Whereas with a scale of 1: 5,000,000 it is impossible to represent different types of crop despite their immense economic interest, with a 1 : 1,000,000 scale and, naturally, larger scales, the main features of agricultural statistics can be inserted on the map against a white background. The potential vegetation of areas at present under cultivation, which it is often difficult to recognize, is given in a panel in ecological colours adjoined to the map. The limits of the various units shown there are indicated on the main map.

With such scales as these, the map is both a botanical and an agricultural one, which makes it more interesting for non-specialists in botany. Various examples have been published. (See, for instance, the 1 : 1,000,000 map of India and the 1 : 200,000 map of France.)

When the scale enables various types of cultivation to be represented, each symbol has a statistical value. For instance, a symbol for vineyards represents 10 hectares (1 ha = 2.471 acres); an olive-grove symbol, 10,000, 1,000 or 100 trees, depending on its size; and so on.

The use of letters enables the percentage of each type of soil utilization in a given administrative area (e.g. a canton) to be indicated. All this depends on the scale, but the system can be used for scales of 1 : 1,000,000 and upwards.

The advantage of the white background is that coloured symbols can be used in relation to the ecology of the type of vegetation represented.

Vegetation symbols

	Evergreens	Deciduous
TREES		
Broad-leaved		
Needle-leaved		
Higher than 50 m		
Coniferous, e.g. pine trees		
Bottle trees		
On termitaries		
Microphyllous		
Thorn trees		
Palm-trees		
SHRUBS		
Broad-leaved		
High scrubs		
Dwarf, dry, eventually deciduous shrubs		
On termitaries		
Microphyllous		
Succulents		
Bamboos		
Shrubs		
Dwarf palm-trees		
Dwarf scrubs		
Cushion dwarf scrubs		
HERBACEOUS VEGETATION		
Perennial forbs		
Ephemeral, episodical forbs		
Ferns		
Lichens		
Moss		
Salicornia		
Creeping vegetation		
Aquatic, floating, rooted vegetation		
Swamps		
Bogs		
Dunes		

Classification internationale et cartographie de la végétation

Table des matières

Introduction

La présente classification a été préparée par le Comité permanent de l'Unesco pour la classification et la cartographie de la végétation du monde. Son but est de fournir un cadre complet pour les catégories les plus importantes à utiliser sur les cartes de la végétation aux échelles de 1/1 000 000 et plus petites [1].

D'abord prévue pour la préparation de nouvelles cartes, la classification peut être utilisée pour transposer dans le nouveau système les cartes déjà existantes.

Le comité a révisé les classifications existantes et conclu qu'elles n'étaient pas entièrement appropriées aux buts présents, bien qu'elles aient influencé la pensée et les décisions du comité. En tant que classification, le système est nécessairement artificiel. L'ordre des unités n'implique pas de relations écologiques et sociologiques.

Les catégories de cette classification sont des unités de végétation comprenant à la fois les formations zonales, les formations azonales ou modifiées les plus importantes et les plus étendues. Il est difficile d'utiliser la composition floristique comme base générale d'une classification pour le monde, car des formations équivalentes en diverses parties du monde proviennent de flores différentes, bien que, dans des régions particulières, la composition floristique joue un rôle important dans la définition et la distinction des communautés.

La physionomie et la structure de la végétation constituent le meilleur ensemble de variables pour une comparaison à l'échelle mondiale. Malheureusement les corrélations ne sont pas toujours claires entre ces deux variables et les habitats écologiques ou les environnements. Pour cette raison, les termes se rapportant au climat, au sol et aux formes de paysage sont inclus dans les dénominations, et parfois dans les définitions, aidant ainsi à identifier certaines unités. Dans la préparation de la classification, il était par conséquent nécessaire d'être pragmatique plutôt que strictement systématique, afin de former des unités avec des noms et des définitions qui soient courts, pleins de sens et descriptifs.

Quoique la plupart des unités soient définies physionomiquement, toutes indiquent les conditions d'environnement. La physionomie de nombreuses prairies ne permet pas à un cartographe de reconnaître leurs affinités de latitude. Indiquer ces affinités dans la classe de formations V de la présente classification physionomique entraînerait donc une multiplication excessive des doubles emplois et répétitions. Toutefois, les auteurs sont invités à employer des termes comme "tropical", "tempéré", etc., à propos des communautés de plantes herbacées, quand ils établissent les légendes des cartes et chaque fois que des renseignements complémentaires de ce genre aideront le lecteur de la carte de la végétation à identifier les catégories dont il s'agit.

L'une des questions qui revient le plus fréquemment au sujet de la classification Unesco de la végétation concerne les aspects dynamiques de la végétation et la place des diverses étapes de la

1. Le système peut aussi être appliqué à des échelles plus grandes en lui ajoutant d'autres subdivisions. Quelques subdivisions plus fines sont introduites à titre d'exemple.

série conduisant au climax. Il est bien évident que ces étapes ne représentent pas toutes des conditions climaciques. Là encore, comme en ce qui concerne l'utilisation de renseignements complémentaires sur le milieu, le comité a été guidé par des principes pragmatiques. A l'échelle de 1/1 000 000 ou à une échelle moindre, il n'est pas possible de représenter les détails qui sont souvent nécessaires pour identifier les communautés climaciques. L'interprétation de la végétation du point de vue du climax est souvent extrêmement difficile, sinon impossible, surtout en ce qui concerne les communautés herbacées. En conséquence, la classification Unesco est fondée sur des types de végétation climacique lorsque cela présente un intérêt pratique. En outre, elle comprend aussi les conditions quasi climaciques et les types de végétation "semi-naturelle", tels qu'ils se présentent au moment de l'observation. Cela donne une certaine latitude au cartographe et lui permet de représenter de nombreux types de végétation dont l'étape dans la série est incertaine, qui sont susceptibles d'être diversement interprétés ou qu'il est actuellement impossible de déterminer. Il n'est pas tenu compte des cultures proprement dites, appelées aussi végétation messicole [1]. Les lecteurs des cartes de végétation fondées sur la classification de l'Unesco ne doivent pas s'attendre à y trouver des champs de blé, des vignobles, des bananeraies ou des rizières [2]. Ces indications sortiraient du cadre de la présente classification.

La classification Unesco de la végétation peut donc être décrite comme ayant un caractère essentiellement physionomico-structurel, avec en outre des renseignements d'ordre écologique intégrés à ses diverses catégories et applicables à la végétation naturelle ou semi-naturelle. Tout en étant relativement complète, elle est conçue de manière à permettre l'adjonction d'autres catégories chaque fois que cela est souhaitable.

1. A. W. Küchler, "Natural and cultural vegetation", *Professional geographer*, vol. 21, p. 383-385.
2. Il est bien évident qu'il serait difficile d'indiquer le climax pour les parties cultivées, car ce serait souvent une forêt, par exemple, mais les cultures occupent à la surface de la terre une place considérable et, pour la cartographie, il faut bien en tenir compte. Si l'on veut s'en tenir aux caractères physionomiques, les céréales seront des herbes, les betteraves et les pommes de terre seront des forbes, la vigne et les arbres fruitiers seront des fruticées. En annexe, p. 64, sera indiquée la méthode utilisée pour la représentation graphique des cultures. (Note de H. Gaussen.)

Classification

Dans la classification Unesco, les unités de rangs différents sont distinguées par des lettres et des chiffres de la façon suivante:

I, II, etc. = CLASSES DE FORMATIONS
A, B, etc. = *SOUS-CLASSES DE FORMATIONS*
1, 2, etc. = GROUPES DE FORMATIONS
a, b, etc. = *Formations*
(1), (2), etc. = Sous-formations
(a), (b), etc. = Divisions suivantes

I FORÊT DENSE
Formée d'arbres d'une taille supérieure à 5 mètres avec des cimes jointives [1].

I.A *FORÊT SURTOUT SEMPERVIRENTE.* Forêt dont la strate arborée n'est jamais défoliée, bien que des arbres isolés puissent perdre leurs feuilles.

I.A.1 FORÊT OMBROPHILE TROPICALE. (Conventionnellement appelée "forêts tropicales humides".) Principalement constituée d'arbres sempervirents, essentiellement avec des bourgeons peu ou pas protégés, et ne résistant ni au froid ni à la sécheresse. Vraiment sempervirente, c'est-à-dire forêt dont la strate arborée reste verte toute l'année, mais où des individus isolés peuvent défolier pendant quelques semaines seulement et pas tous en même temps. Les feuilles de nombreuses espèces sont acuminées.

1 I.A.1a *Forêt ombrophile tropicale de basse altitude.* Habituellement formée de nombreuses espèces d'arbres à croissance rapide, un grand nombre d'entre eux dépassant 50 mètres de haut [2], généralement à écorce lisse, souvent épaisse, avec quelques contreforts. Des arbres émergeants ou tout au moins une strate arborée très inégale souvent présents. Strate inférieure très clairsemée et principalement composée par les plantules des arbres.

1. Dans les stades de reproduction ou dans les cas de formations secondaires immaturées, la taille peut être inférieure à 5 mètres, mais les individus n'ont qu'un seul tronc (ce sont de vrais arbres, non des arbustes). La taille limite peut être de 3 mètres dans les conditions subpolaires et de 2 à 10 mètres dans les conditions tropicales.
2. Les limites de hauteur ne sont qu'un indice général et non pas un critère absolu.

Palmiers et autres arbres en touffe habituellement rares, lianes presque absentes, excepté les pseudo-lianes (c'est-à-dire plantes se formant sur les branches des arbres, puis s'enracinant dans le sol, ou vice versa).

Lichens en croûte et algues vertes sont les seules formes de vie épiphytique constamment présentes; épiphytes vasculaires généralement peu abondantes; abondantes seulement dans les situations excessivement humides (ex.: Sumatra, Atrato Valley [Colombie], etc.).

2 I.A.1b *Forêt ombrophile tropicale submontagnarde.* Absence nette d'arbres émergeants et strate relativement uniforme. Dans la strate inférieure, forbes communes. Épiphytes vasculaires et pseudo-lianes abondantes (ex.: versant Atlantique du Costa Rica).

I.A.1c *Forêt ombrophile tropicale montagnarde.* Plantes vasculaires et autres épiphytes abondantes. Arbres généralement inférieurs à 50 mètres; cimes descendant relativement bas le long du tronc. Écorce souvent plus ou moins rugueuse. Strate inférieure abondante, souvent représentée par des micro- et nano-phanérophytes en rosette (ex.: fougères arborescentes ou petits palmiers); couverture du sol riche en herbacées et cryptogames hygromorphes (ex.: Sierra de Talamanca [Costa Rica]).

3 I.A.1c(1) Feuillus dominant.

4 I.A.1c(2) Résineux.

5 I.A.1c(3) Microphylles.

6 I.A.1c(4) Bambous, riche en herbacées arborées qui remplacent largement les micro- ou nano-phanérophytes en touffe.

7 I.A.1d *Forêt ombrophile tropicale subalpine.* (Forêt de brouillard et forêt claire non comprises. Considérée comme existante par quelques chercheurs, mais probablement pas importante. Définition nécessaire.)

I.A.1e *Forêt ombrophile tropicale de brouillard.* Les cimes, branches et troncs aussi bien que les lianes, chargés d'épiphytes, principalement des bryophytes chamæphytiques. Sol également couvert de chamæphytes hygromorphes (ex.: *Selaginella* et fougères herbacées). Arbres souvent noueux, à écorce rugueuse et dépassant rarement 20 mètres de haut (ex.: Blue Mountains [Jamaïque]).

8 I.A.1e(1) Feuillus, forme la plus commune.

9 I.A.1e(2) Résineux.

10 I.A.1e(3) Microphylles.

I.A.1f *Forêt ombrophile tropicale sur alluvions.* Semblable à la forêt submontagnarde (I.A.1b), mais plus riche en palmiers et strate inférieure plus diversifiée, particulièrement en grandes forbes (ex.: *Musaceae*); contreforts fréquents (ex.: bassin de l'Amazone).

11 I.A.1f(1) Riveraine (sur les rives boisées les plus basses, fréquemment inondées), principalement dominée par des arbres à croissance rapide; strate inférieure herbacée presque absente, épiphytes extrêmement rares, pauvre en espèces (ex.: igapó amazonien).

12 I.A.1f(2) Occasionnellement inondée sur terrasses relativement sèches accompagnant les rivières actives. Plus d'épiphytes que dans (1) et (3), beaucoup de lianes.

13 I.A.f(3) En terrain saisonnièrement gorgé d'eau (le long des cours d'eau les plus bas, aux endroits où l'eau s'accumule en grandes flaques pour plusieurs mois, souvent derrière de basses digues naturelles); arbres possédant fréquemment des racines-échasses; densité de la strate arborée non uniforme; il est de règle que la strate inférieure soit pauvre, sauf aux emplacements plus ouverts (ex.: várzea amazonienne).

I.A.1g *Forêt ombrophile tropicale marécageuse.* Non pas le long des rivières, mais sur sol mouilleux, pouvant être approvisionné en eau fraîche et en eau saumâtre. Semblable à la forêt alluviale (I.A.1f) mais en général plus pauvre en espèces arborées. Nombreux arbres avec contreforts ou pneumatophores; la plupart dépassant 20 mètres de haut (ex.: Sumatra oriental).

14 I.A.1g(1) Feuillus, dominance des plantes dicotylédones.

15 I.A.1g(2) Dominance des palmiers, mais arbres feuillus dans la strate inférieure (ex.: marécages de *Raphia taedigera* du Costa Rica).

16 I.A.1h *Forêt sempervirente tropicale sur tourbière.* (Avec dépôts organiques de surface.) Pauvre en espèces arborées, avec strate supérieure formée d'arbres dont la taille est élevée en bordure de tourbière mais décroît vers le centre. Les arbres ont souvent de faibles diamètres et sont communément pourvus de pneumatophores ou de racines-échasses.

I.A.2 FORÊT SEMPERVIRENTE SAISONNIÈRE TROPICALE ET SUBTROPICALE. Principalement formée d'arbres sempervirents dont les bourgeons sont en partie protégés. La réduction du feuillage pendant la saison sèche est notable, souvent sous forme de chute partielle des feuilles. Fait transition entre I.A.1 et I.A.3. Subdivisions a-c très comparables à celles de la forêt ombrophile tropicale (I.A.1).

17 I.A.2a *Forêt sempervirente saisonnière tropicale (ou subtropicale) de basse altitude.*

I.A.2b *Forêt sempervirente saisonnière tropicale (ou subtropicale) submontagnarde.*

18 I.A.2b(1) Feuillus.

19 I.A.2b(2) Résineux.

20 I.A.2c *Forêt sempervirente saisonnière tropicale (ou subtropicale) montagnarde.* Diffère de I.A.1c par l'absence de fougères arborescentes; par contre, buissons sempervirents plus fréquents.

21 I.A.2d *Forêt sempervirente sèche tropicale (ou subtropicale) subalpine.* Ressemble physionomiquement à la forêt sclérophylle sempervirente sèche à pluies d'hiver (I.A.8a); existe normalement au-dessus de la forêt de brouillard (I.A.1e). Arbres surtout sclérophylles sempervirents, inférieurs à 20 mètres, avec peu ou pas de strate inférieure (si elle n'a pas été ouverte par l'homme). Pauvre en lianes et épiphytes, lichens exceptés.

I.A.3 FORÊT SEMI-DÉCIDUE TROPICALE ET SUBTROPICALE. Arbres de la strate supérieure en majeure partie décidus en saison sèche; nombreux arbres de la strate intermédiaire et arbustes du sous-bois sempervirents et plus ou moins sclérophylles. Pourtant les plantes ligneuses sempervirentes et décidues ne forment pas toujours des strates distinctes. Elles peuvent exister en mélange dans la même strate, ou encore les arbrisseaux peuvent être essentiellement décidus et les arbres sempervirents. Presque tous les arbres ont des bourgeons protégés; feuilles acuminées. Les arbres présentent une écorce rugueuse, excepté quelques arbres bouteilles qui peuvent être présents.

22 I.A.3a *Forêt semi-décidue tropicale ou subtropicale de basse altitude.* Les arbres les plus hauts sont souvent des arbres bouteilles (ex.: *Ceiba*). Pratiquement pas d'épiphytes. Strate inférieure composée de plantules d'arbres et d'arbrisseaux vraiment ligneux. Des plantes succulentes peuvent être présentes (ex.: des cactées cespiteuses à tige grêle). Il y a occasionnellement des lianes à la fois thérophytes et hémicryptophytes. Il peut y avoir aussi une couche d'herbe clairsemée principalement formée d'hémicryptophytes graminoïdes et de forbes.

23 I.A.3b *Forêt semi-décidue tropicale (ou subtropicale) montagnarde ou de brouillard.* Analogue à la forêt semi-décidue de basse altitude (I.A.3a), mais strate arborée plus basse et couverte d'épiphytes xérophytiques (ex.: *Tillandsia usneoides*). Dans le groupe I.A.3, semi-décidu, une formation submontagnarde ne peut être nettement distinguée.

I.A.4 FORÊT OMBROPHILE SUBTROPICALE. Ne se présente que localement et en peuplements fragmentaires car le climat subtropical est typiquement un climat avec une saison sèche. Lorsque la forêt ombrophile subtropicale existe, comme par exemple au Queensland (Australie) et à Taïwan, elle passe presque insensiblement à la forêt ombrophile tropicale. La strate inférieure peut comporter quelques arbustes. La forêt ombrophile subtropicale ne devrait pourtant pas être confondue avec la forêt ombrophile de montagne, qui existe sous un climat à température moyenne annuelle comparable, mais avec des différences de température moins prononcées entre l'été et l'hiver. Par conséquent, les rythmes saisonniers sont plus évidents dans toutes les forêts subtropicales, même dans les forêts ombrophiles.

La forêt ombrophile subtropicale est physionomiquement plus proche des formes tropicales que des formes tempérées. Ses subdivisions correspondent donc plus ou moins à celles de la forêt ombrophile tropicale (I.A.1a à h).

24 I.A.4a

25 I.A.4b

26 I.A.4c(1)
27 I.A.4c(2)
28 I.A.4c(3)
29 I.A.4c(4)

30 I.A.4d

31 I.A.4e(1)
32 I.A.4e(2)
33 I.A.4e(3)

34 I.A.4f(1)
35 I.A.4f(2)
36 I.A.4f(3)

37 I.A.4g(1)
38 I.A.4g(2)

39 I.A.4h

40 I.A.5 FORÊT DE MANGROVE. Existe seulement dans la zone intercotidale des régions tropicales et subtropicales. Presque entièrement composée d'arbres et d'arbustes feuillus sclérophylles sempervirents avec des racines-échasses ou des pneumatophores. Épiphytes généralement rares, excepté les lichens se développant sur les branches et les algues sur les parties basses des troncs (subdivisions possibles; des transitions existent avec la forêt ombrophile marécageuse tropicale [I.A.1g]; ex.: côtes de Borneo, Nouvelle-Guinée, etc.).

I.A.6 FORÊT SEMPERVIRENTE OMBROPHILE TEMPÉRÉE ET SUBPOLAIRE. Existe seulement en climat extrêmement océanique, sans gel, dans l'hémisphère sud, principalement au Chili. Principalement formée d'arbres et arbustes hémisclérophylles vraiment sempervirents. Riche en thalloépiphytes et en fougères herbacées terrestres.

I.A.6a *Forêt sempervirente ombrophile tempérée de feuillus*. Quelques épiphytes vasculaires et lianes sont présentes; hauteur généralement au-dessus de 10 mètres.

41 I.A.6a(1) Feuillus principalement (ex.: forêts de *Nothofagus* de Nouvelle-Zélande).

42 I.A.6a(2) Avec mélange de résineux;

43 I.A.6a(3) Principalement résineux ou arbres à feuilles-écailles (ex.: forêts de *Podocarpus* de Nouvelle-Zélande).

44 I.A.6b *Forêt sempervirente ombrophile tempérée sur alluvions*. Riche en forbes (ex.: Nouvelle-Zélande occidentale).

I.A.6c — *Forêt sempervirente ombrophile marécageuse tempérée.*

45 I.A.6c(1) — A feuilles en aiguilles ou en écailles. Forêt dense, haute (jusqu'à 50 mètres et plus), à feuilles en écailles, de basse altitude. Contreforts fréquents. Couverture du sol (continue à claire) formée de graminoïdes (surtout *Carex*) et de forbes (surtout fougères). Riche en épiphytes vasculaires et bryoïdes; quelques lianes (ex.: communautés de *Podocarpus dacrydioides* de Nouvelle-Zélande).

46 I.A.6c(2) — A feuilles larges. Forêt haute (jusqu'à 50 mètres et plus) latifoliée de basse altitude. Arbres en peuplement dense avec des cimes jointives, mais la voûte laisse passer beaucoup de lumière. Synusie buissonneuse assez ouverte. Épiphytes absentes dans la voûte (ex.: forêts à *Eucalyptus ovata* du Victoria).

47 I.A.6d — *Forêt sempervirente ombrophile subpolaire.* Diffère de la forêt feuillue tempérée (I.A.6a) par le manque d'épiphytes vasculaires et par la hauteur plus réduite du peuplement (généralement inférieure à 10 mètres). Taille des feuilles généralement réduite elle aussi (ex.: forêts de hêtres de Nouvelle-Zélande).

I.A.7 — FORÊT SEMPERVIRENTE SAISONNIÈRE TEMPÉRÉE DE FEUILLUS (AVEC PLUIES D'ÉTÉ SUFFISANTES). Comprend principalement des arbres et des arbustes hémisclérophylles sempervirents. Riche en chamæphytes et hémicryptophytes herbacées dans la strate inférieure. Très peu ou pas d'épiphytes vasculaires et lianes. Passe progressivement à la forêt ombrophile subtropicale (I.A.4) ou tempérée (I.A.6) ou à la forêt sclérophylle de feuillus sempervirents à pluies d'hiver (I.A.8). Inclut probablement les types subpolaires.

(Des subdivisions comparables à celles des forêts tropicales et subtropicales, I.A.2a à d, sont possibles.)

48 I.A.7a

49 I.A.7b(1)

50 I.A.7b(2)

51 I.A.7c

52 I.A.7d

I.A.8 — FORÊT SEMPERVIRENTE DE FEUILLUS SCLÉROPHYLLES A PLUIES D'HIVER. (Souvent considérée comme étant méditerranéenne, mais présente aussi dans le Sud-Ouest australien, au Chili, etc. Climat à été sec prononcé.) Principalement formée d'arbres et d'arbustes sclérophylles sempervirents, dont la plupart présentent une écorce rugueuse. Strate inférieure herbacée pratiquement absente. Pas d'épiphytes vasculaires et très peu d'épiphytes cryptogamiques, mais présence de lianes ligneuses sempervirentes.

53 I.A.8a — *Forêt sempervirente sclérophylle à pluies d'hiver, de basse altitude (incluant la forêt submontagnarde).* Composée d'eucalyptus géants (ex.: *Eucalyptus regnans* à Victoria et *E. diversicolor* en Australie occidentale). Dominée par des arbres de plus de 50 mètres de haut.

54 I.A.8b — Amplement décrit dans I.A.8, mais arbres de moins de 50 mètres de haut (ex.: forêts de chênes verts de Californie).

55 I.A.8c — La forêt sur alluvions et sur marécage de ce type existe peut-être, mais n'est pas suffisamment connue.

I.A.9 — FORÊT SEMPERVIRENTE TROPICALE ET SUBTROPICALE DE CONIFÈRES. Formée principalement d'arbres sempervirents à aiguilles ou à écailles. Des feuillus peuvent y être mélangés. Les épiphytes vasculaires et les lianes en sont pratiquement absentes.

56 I.A.9a *Forêt sempervirente tropicale et subtropicale de conifères, de basse altitude et submontagnarde* (ex.: forêt de pins du Honduras et du Nicaragua).

57 I.A.9b *Forêt sempervirente tropicale et subtropicale des régions de montagne et subalpines* (ex.: forêt de pins des Philippines et du Mexique méridional).

I.A.10 FORÊT SEMPERVIRENTE TEMPÉRÉE ET SUBPOLAIRE DE CONIFÈRES. Formée principalement d'arbres sempervirents à aiguilles ou à écailles, mais des feuillus peuvent y être mélangés. Les épiphytes vasculaires et les lianes en sont pratiquement absentes.

58 I.A.10a *Forêt sempervirente de conifères géants.* Dominée par des arbres dépassant 50 mètres (ex.: forêts de *Sequoia* et *Pseudotsuga* dans la région Pacifique de l'Amérique du Nord).

I.A.10b *Forêt sempervirente de conifères avec houppiers arrondis.* Dominée par des arbres de 45 à 50 mètres de haut, avec des houppiers plus ou moins larges irrégulièrement arrondis (ex.: *Pinus* spp.).

59 I.A.10b(1) Avec strate inférieure sempervirente sclérophylle.

60 I.A.10b(2) Sans strate inférieure sempervirente sclérophylle.

61 I.A.10c *Forêt sempervirente de conifères avec houppiers coniques.* Dominée par des arbres de 45 à 50 mètres de haut (exceptionnellement plus hauts) à cimes plus ou moins coniques, comme la plupart des *Picea* et *Abies* (ex.: forêts de sapins de Douglas [*Pseudotsuga Douglasii*] de Californie).

62 I.A.10d *Forêt sempervirente de conifères avec houppiers cylindriques (boréale).* Comparable à I.A.1c mais les cimes ont des branches très courtes et sont par conséquent très étroites, cylindro-coniques.

I.B *FORÊT SURTOUT DÉCIDUE.* La majorité des arbres perdent leur feuillage en même temps pendant la saison défavorable.

I.B.1 FORÊT DÉCIDUE EN SAISON SÈCHE (TROPICALE OU SUBTROPICALE). Saison défavorable caractérisée par la sécheresse; dans la plupart des cas sécheresse hivernale. Chute des feuilles régulièrement tous les ans. La plupart des arbres ont une écorce relativement épaisse et fissurée.

63 I.B.1a *Forêt décidue en saison sèche, de basse altitude (et submontagnarde).* Dans toutes les strates, pratiquement pas de plantes sempervirentes, excepté quelques plantes succulentes. Présence occasionnelle de lianes ligneuses et herbacées, ainsi que d'arbres bouteilles décidus. Couverture du sol essentiellement herbacée mais claire (hémicryptophytes, surtout des graminées, des géophytes et quelques thérophytes) [ex.: forêts décidues d'arbres feuillus du nord-ouest du Costa Rica].

64 I.B.1b *Forêt décidue en saison sèche, de montagne (et de brouillard).* Quelques espèces sempervirentes dans le sous-bois. Présence ou abondance d'épiphytes résistant à la sécheresse, souvent de formes barbues (ex.: *Usnea* ou *Tillandsia usneoides*). Cette formation n'est pas fréquente, mais elle est bien développée, par exemple dans le nord du Pérou.

I.B.2 FORÊT DÉCIDUE EN SAISON FROIDE, MÉLANGÉE D'ARBRES (OU D'ARBUSTES) SEMPERVIRENTS. La mauvaise saison est principalement caractérisée par le gel d'hiver. Dominance des feuillus décidus, mais présence d'espèces sempervirentes dans la strate supérieure ou constituant le sous-bois. Rareté ou absence de plantes grimpantes et d'épiphytes vasculaires.

65 I.B.2a *Forêt décidue en saison froide avec feuillus sempervirents et plantes grimpantes* (ex.: *Ilex aquifolium* et *Hedera helix* en Europe occidentale). Riche en épiphytes cryptogamiques, y compris les mousses. Les épiphytes vasculaires peuvent être présents à la base des troncs. Lianes parfois communes dans les plaines d'inondation.

66 I.B.2b *Forêt décidue en saison froide avec résineux sempervirents* (ex.: forêts d'érables et de pruches [*Tsuga*] de New York).

I.B.3 FORÊT DÉCIDUE EN SAISON FROIDE SANS ARBRES SEMPERVIRENTS. Dominance absolue d'arbres décidus. Présence possible de chamæphytes sempervirentes et de quelques nanophanérophytes sempervirentes. Plantes grimpantes insignifiantes (mais peuvent être communes dans les plaines d'inondation); épiphytes vasculaires absentes (excepté occasionnellement à la base des arbres); thalloépiphytes toujours présentes, en particulier des lichens.

67 I.B.3a *Forêt décidue en saison froide, tempérée, de basse altitude et submontagnarde.* Arbres atteignant 50 mètres de haut (ex.: forêt mésophytique mélangée, aux États-Unis). Épiphytes, essentiellement algues et lichens en croûte.

I.B.3b *Forêt décidue en saison froide de montagne ou boréale.* (Incluant la forêt de basse altitude ou submontagnarde, dans les positions topographiques de forte humidité atmosphérique.) Épiphytes: lichens foliacés et frutiqueux; bryophytes. Arbres jusqu'à 50 mètres de haut, mais ne dépassant pas 30 mètres dans les forêts de montagne ou boréales.

68 I.B.3b(1) Essentiellement feuillus décidus.

69 I.B.3b(2) Essentiellement résineux décidus (ex.: *Larix* de Sibérie).

70 I.B.3b(3) Mélange de feuillus et de résineux décidus.

I.B.3c *Forêt décidue en saison froide subalpine ou subpolaire.* Diffère de la forêt décidue en saison froide de basse altitude et de montagne (I.B.3a et b) par la hauteur de la strate arborée, qui est nettement réduite (ne dépassant pas 20 mètres). Tronc souvent noueux. Épiphytes comme dans (b), mais en général plus abondantes. Passe souvent progressivement à la forêt claire (voir classe II).

71 I.B.3c(1) Avec strate inférieure essentiellement constituée d'hémicryptophytes.

72 I.B.3c(2) Avec strate inférieure essentiellement constituée de chamæphytes. Peut fusionner avec des forêts mélangées de conifères.

I.B.3d *Forêt décidue en saison froide sur alluvions.* Inondée par les rivières, par conséquent plus humide et plus riche en éléments nutritifs que la forêt décidue de saison froide de basse altitude (I.B.3a). Arbres et arbustes à croissance rapide et herbacées vigoureuses dans la strate inférieure.

73 I.B.3d(1) Occasionnellement inondée; physionomiquement semblable à I.B.3a, avec grands arbres et abondante strate inférieure frutescente et macrophylle.

74 I.B.3d(2) Régulièrement inondée; arbres moins hauts et moins denses que dans I.B.3a, mais dans la strate inférieure herbacées abondantes et hautes (en Eurasie des espèces de *Salix* ou d'*Alnus* dominent fréquemment).

I.B.3e *Forêt décidue en saison froide de marécage ou de tourbière.* Inondée jusqu'à la fin du printemps ou au début de l'été; sol de surface organique. Relativement pauvre en espèces arborées. Couverture vivante plutôt continue, constituée de formes biologiques variées. [Mêmes subdivisions que pour la forêt décidue de saison froide boréale (I.B.3b).]

75 I.B.3e(1)

76 I.B.3e(2)

77 I.B.3e(3)

I.C *FORÊT EXTRÊMEMENT XÉROMORPHE.* Peuplements denses de phanérophytes xéromorphes, telles que les arbres bouteilles, arbres en touffe à feuilles succulentes et plantes succulentes dressées. Strate inférieure à arbustes présentant des adaptations xéromorphiques comparables, chamæphytes succulentes et hémicryptophytes herbacées, géophytes et thérophytes. Souvent passage progressif à la forêt claire (voir classe II).

78 I.C.1 FORÊT EXTRÊMEMENT XÉROMORPHE A DOMINANCE SCLÉROPHYLLE. Combinaison de formes biologiques comme ci-dessus, excepté pour la prédominance des arbres sclérophylles, dont la plupart ont des tiges à base bulbeuse, largement enfoncées dans le sol (xylopodes).

I.C.2 FORÊT D'ÉPINEUX. Prédominance des espèces avec appendices épineux.

79 I.C.2a *Forêt mélangée d'épineux sempervirents et décidus.*

80 I.C.2b *Forêt d'épineux exclusivement décidus.*

81 I.C.3 FORÊT A DOMINANCE DE SUCCULENTS. Succulents fréquemment unitiges (scapeux) et en touffe (cespiteux); mais les autres xérophanérophytes sont aussi présentes.

II FORÊT CLAIRE (peuplement d'arbres ouvert)

Formée d'arbres d'au moins 5 mètres de haut, la plupart des cimes ne se touchant pas les unes les autres, mais couvrant au moins 40% de la surface. Présence possible d'une synusie herbacée. Voir le groupe de formations V.A.1 si les arbres couvrent moins de 40% de la surface du sol et s'il y a une synusie herbacée. La limite de 40% de couverture du sol est commode parce qu'on peut l'évaluer facilement au cours des travaux sur le terrain: lorsque la couverture arborée est de 40%, la distance entre deux houppiers est égale au rayon moyen d'un houppier.

II.A *FORÊT CLAIRE SURTOUT SEMPERVIRENTE*, (c'est-à-dire sempervirente comme défini en I.A.).

82 II.A.1 FORÊT CLAIRE SEMPERVIRENTE DE FEUILLUS. Arbres et arbustes principalement sclérophylles; pas d'épiphytes.

II.A.2 FORÊT CLAIRE SEMPERVIRENTE DE RÉSINEUX. Principalement aiguilles ou feuilles-écailles. Les houppiers de beaucoup d'arbres descendent jusqu'à la base des troncs ou sont, tout au moins, très branchus.

II.A.2a *Forêt claire sempervirente de résineux (conifères) avec houppier arrondi* (ex.: *Pinus*).
83 II.A.2a(1) Sous-bois sclérophylle sempervirent (méditerranéen);
84 II.A.2a(2) Sans sous-bois sclérophylle sempervirent.

85 II.A.2b *Forêt claire sempervirente de résineux avec prédominance de houppiers coniques* (surtout subalpine).

86 II.A.2c *Forêt claire sempervirente de conifères avec houppiers cylindro-coniques très étroits* (ex.: *Picea* dans les régions boréales).

II.B *FORÊT CLAIRE SURTOUT DÉCIDUE* (voir I.B).

II.B.1 FORÊT CLAIRE EN SAISON SÈCHE. (Subdivisions comparables à celles des forêts.)

87 II.B.1a

88 II.B.1b

II.B.2 FORÊT CLAIRE DÉCIDUE EN SAISON FROIDE AVEC ARBRES SEMPERVIRENTS. (Voir I.B.2.)

89 II.B.2a

90 II.B.2b

II.B.3 FORÊT CLAIRE DÉCIDUE EN SAISON FROIDE SANS ARBRES SEMPERVIRENTS. (Voir I.B.3.) Plus fréquente dans la région subantarctique; ailleurs ne se trouvant que sur marécages ou tourbières.

91 II.B.3a *Forêt claire décidue de feuillus.*

92 II.B.3b *Forêt claire décidue de résineux.*

93 II.B.3c *Forêt claire décidue mélangée (feuillus et résineux).*

II.C *FORÊT CLAIRE EXTRÊMEMENT XÉROMORPHE.* Semblable à I.C, la seule différence étant le mode de peuplement des arbres, ici plus clairsemés. (Subdivisions comme sous I.C.)

94 II.C.1

95 II.C.2a

96 II.C.2b

97 II.C.3

III FRUTICÉE (formations buissonneuses et fourrés)

Surtout composée de phanérophytes ligneuses cespiteuses de 0,5 à 5 mètres de haut [1]. Les subdivisions suivantes pourraient être:

Formations buissonneuses. La plupart des individus ne se touchent pas les uns les autres; présence fréquente d'une strate graminéenne.

Fourrés. Arbustes entremêlés.

III.A *FRUTICÉE SURTOUT SEMPERVIRENTE* (sempervirente au sens de I.A).

III.A.1 BUISSONS (OU FOURRÉS) SEMPERVIRENTS DE FEUILLUS.

98 III.A.1a *Fourrés à petits bambous (ou, moins fréquemment, buissons).* Nano- et micro-phanérophytes graminoïdes rampantes, lignifiées.

99 III.A.1b *Buissons (ou fourrés) sempervirents arbustifs.* Composés de petits arbres et d'arbustes ligneux (ex.: palmiers nains de Méditerranée ou fourrés de fougères arborescentes d'Hawaii).

100 III.A.1c *Fourrés (ou buissons) sempervirents de feuillus hémisclérophylles.* Nano- ou micro-phanérophytes couchées, rampantes, cespiteuses, avec feuilles relativement grandes et tendres (ex.: fourrés subalpins de rhododendrons, ou fourrés denses d'*Hibiscus liliaceus* d'Hawaii).

101 III.A.1d *Buissons (ou fourrés) sempervirents de feuillus sclérophylles.* Dominance d'arbustes sclérophylles et d'arbres immatures (ex.: chaparral ou maquis). Peut fusionner avec la forêt-parc, la prairie ou la lande.

102 III.A.1e *Fourrés (ou buissons) sempervirents suffrutescents.* Peuplements de nano-phanérophytes semi-lignifiées qui, dans les années sèches, peuvent perdre une partie des rameaux de l'année (ex.: lande à *Cistus*) [2].

1. Ne pas confondre avec forêt secondaire en cours de développement (voir note p. 43). Quelquefois les arbustes peuvent atteindre plus de 5 mètres de haut.
2. Occasionnellement hauteur inférieure à 50 cm, donc passant progressivement vers IV.A.1a.

	III.A.2	BUISSONS (OU FOURRÉS) SEMPERVIRENTS A MICROPHYLLES ET RÉSINEUX.
103	III.A.2a	*Fourrés (ou buissons) sempervirents de résineux.* Surtout composés de phanérophytes résineuses rampantes et couchées (ex.: *Pinus mughus*, "Krummholz").
104	III.A.2b	*Buissons (ou fourrés) sempervirents microphylles.* Souvent arbustes éricoïdes (surtout dans les étages subalpins tropicaux).
	III.B	*FRUTICÉE PRINCIPALEMENT DÉCIDUE* (décidue au sens de I.B).
105	III.B.1	FRUTICÉE DÉCIDUE EN SAISON SÈCHE AVEC PLANTES LIGNEUSES SEMPERVIRENTES EN MÉLANGE.
106	III.B.2	FRUTICÉE DÉCIDUE EN SAISON SÈCHE SANS PLANTES LIGNEUSES SEMPERVIRENTES EN MÉLANGE.
	III.B.3	FRUTICÉE DÉCIDUE EN SAISON FROIDE.
107	III.B.3a	*Fourrés (ou buissons) décidus tempérés.* Fruticée plus ou moins dense avec ou sans sous-étage formé seulement de petites herbacées; pauvre en cryptogames.
	III.B.3b	*Fourrés (ou buissons) décidus subalpins ou subpolaires.* Nano-phanérophytes cespiteuses dressées ou prostrées à forte capacité de régénération végétative. En règle générale couverts de neige pendant la moitié de l'année au moins.
108	III.B.3b(1)	Avec strate inférieure essentiellement d'hémicryptophytes, surtout des forbes (ex.: fourrés subalpins d'*Alnus viridis*).
109	III.B.3b(2)	Avec strate inférieure essentiellement de chamæphytes, surtout arbustes nains et lichens frutescents (ex.: les fourrés de *Betula tortuosa* à la limite des arbres des régions polaires).
	III.B.3c	*Buissons (ou fourrés) décidus sur alluvions.* Arbustes à croissance rapide, colonisateurs des berges des rivières et des îles qui sont fortement inondées et de façon assez fréquente, donc avec strate inférieure surtout clairsemée.
110	III.B.3c(1)	Avec feuilles lancéolées (ex.: *Salix*, surtout dans les régions de basse altitude et submontagnardes).
111	III.B.3c(2)	Formes microphylles.
112	III.B.3d	*Buissons (ou fourrés) décidus sur tourbe.* Nano-phanérophytes dressées cespiteuses, avec *Sphagnum* et (ou) d'autres mousses de tourbe.
	III.C	*FORMATION BUISSONNEUSE (SUBDÉSERTIQUE) EXTRÊMEMENT XÉROMORPHE.* Très souvent peuplements clairs d'arbustes, présentant diverses adaptations xérophytiques, telles que: feuilles extrêmement scléromorphes ou fortement réduites, rameaux verts sans feuilles, ou tiges succulentes, etc., certains avec des épines.
	III.C.1	FORMATION BUISSONNEUSE SURTOUT SEMPERVIRENTE SUBDÉSERTIQUE. Dans les années extrêmement sèches quelques feuilles et quelques portions de rameaux peuvent être caduques.
	III.C.1a	*Formation buissonneuse exclusivement sempervirente subdésertique.*
113	III.C.1a(1)	Feuillus, avec dominance des nano-phanérophytes sclérophylles; quelques arbustes à phyllodes sont inclus (ex.: la fruticée à " mulga " d'Australie).
114	III.C.1a(2)	Formes à microphylles ou sans feuilles, mais avec des tiges vertes (ex.: *Retama retam*).
115	III.C.1a(3)	Succulents, dominés par des formes diversement branchues à feuilles succulentes.
	III.C.1b	*Formation buissonneuse subdésertique semi-décidue.* Soit arbustes, facultativement décidus, soit combinaison d'arbustes sempervirents et décidus.

116 III.C.1b(1) Facultativement décidus (ex.: le "bush" salé à *Atriplex* et *Kochia* d'Australie).

117 III.C.1b(2) Mélange de sempervirents et de décidus, transition vers III.C.2.

III.C.2 FORMATION BUISSONNEUSE SUBDÉSERTIQUE DÉCIDUE. Surtout arbustes décidus, souvent avec quelques sempervirents.

118 III.C.2a *Formation buissonneuse subdésertique décidue sans succulents.*

119 III.C.2b *Formation buissonneuse subdésertique décidue avec succulents.*

IV FRUTICÉE NAINE ET FORMATIONS ANALOGUES

Excédant rarement 50 cm de haut (parfois appelées landes).
En fonction de la densité de la couverture des buissons, on distingue:
Les fourrés nains: branches entremêlées;
Les formations buissonneuses naines: buissons plus ou moins isolés ou en touffes;
Les formations cryptogamiques avec des buissons nains: surface densément recouverte de mousses ou lichens (thallo-chamæphytes); buissons nains en touffes ou isolés. Dans le cas des tourbières, des formations graminoïdes peuvent prédominer localement.

IV.A *FRUTICÉE SURTOUT SEMPERVIRENTE.* La plupart des buissons sont sempervirents.

IV.A.1 FOURRÉS NAINS SEMPERVIRENTS. Prédominance d'une couverture buissonneuse dense et fermée dans le paysage (lande naine au sens propre).

120 IV.A.1a *Fourrés nains sempervirents cespiteux.* Les branches dressées sont pour la plupart couvertes de lichens foliacés. Sur le sol, mousses pulvinées, lichens frutescents, ou formes biologiques herbacées peuvent tenir une certaine place (ex.: la lande à *Calluna*).

121 IV.A.1b *Fourrés nains sempervirents rampants ou en brosse.* La plupart des branches rampent sur le sol. Combinés de façon variable avec des thallo-chamæphytes dont les branches peuvent être enfouies (ex.: la lande à *Loiseleuria*).

IV.A.2 BUISSONS NAINS SEMPERVIRENTS. Formation d'arbustes nains ouverte ou assez lâche.

122 IV.A.2a *Buissons sempervirents en coussin.* Touffes plus ou moins isolées d'arbustes nains, formant des coussins denses, souvent munis d'épines (ex.: les landes porcs-épics à *Astragalus* et *Acantholimon* des montagnes de l'Est méditerranéen).

IV.A.3 MÉLANGE DE FOURRÉS NAINS SEMPERVIRENTS ET DE FORMATIONS HERBACÉES. Peuplements plus ou moins ouverts de plantes suffrutescentes sempervirentes ou de chamæphytes herbacées, hémicryptophytes variées, géophytes, etc.

123 IV.A.3a *Fourrés nains exclusivement sempervirents et formation herbeuse mélangée* (ex.: lande à *Nardus* et *Calluna*).

124 IV.A.3b *Fourrés nains partiellement sempervirents et formation herbeuse mélangée.* De nombreux individus perdent une partie de leurs rameaux de l'année pendant la saison sèche (ex.: phrygane de Grèce).

IV.B *FRUTICÉE NAINE SURTOUT DÉCIDUE.* Analogue à IV.A, mais surtout formée d'espèces décidues.

125 IV.B.1 FOURRÉS NAINS OU BUISSONS NAINS FACULTATIVEMENT DÉCIDUS EN SAISON SÈCHE. Caduques seulement dans les années extrêmes.

IV.B.2 FOURRÉS NAINS OU BUISSONS NAINS (OBLIGATOIREMENT) DÉCIDUS EN SAISON SÈCHE. Peuplements de fourrés nains très fermés qui perdent tout ou partie de leurs feuilles en saison sèche.

126 IV.B.2a *Fourrés nains, décidus en saison sèche, cespiteux.* Correspondant à IV.A.1a.

127 IV.B.2b *Fourrés nains, décidus en saison sèche, rampants ou en brosse.* Correspondant à IV.A.1b.

128 IV.B.2c *Buissons nains, décidus en saison sèche, en coussin.* Correspondant à A.2a.

129 IV.B.2d *Buissons nains, décidus en saison sèche, en mosaïque (ou mélangés).* Arbustes nains décidus ou sempervirents, hémicryptophytes cespiteuses, chamæphytes succulentes et autres formes biologiques mêlées de diverses manières.

IV.B.3 FOURRÉS NAINS (OU BUISSONS NAINS) DÉCIDUS EN SAISON FROIDE. Physionomiquement semblables à IV.B.2, mais perdant leurs feuilles au commencement de la saison froide. Habituellement plus riches en chamæphytes cryptogamiques.

130 IV.B.3a (Subdivisions comme dans IV.B.2. Transitions vers IV.D et E possibles.)

131 IV.B.3b

132 IV.B.3c

133 IV.B.3d

IV.C *BUISSONS NAINS EXTRÊMEMENT XÉROMORPHES.* Formation plus ou moins ouverte formée d'arbustes nains, succulents, géophytes, thérophytes et autres formes biologiques adaptées à survivre à (ou à éviter) une longue saison sèche. Surtout subdésertique.

134 IV.C.1a(1) (Subdivisions comme III.C, buissons extrêmement xéromorphes.)
135 IV.C.1a(2)
136 IV.C.1a(3)

137 IV.C.1b(1)
138 IV.C.1b(2)

139 IV.C.2a

140 IV.C.2b

IV.D *TOUNDRA.* Formation basse à croissance lente, principalement de buissons nains, de graminées, et de cryptogames, au-delà de la limite subpolaire des arbres, montrant souvent une disposition des plantes causée par les mouvements du sol dus au gel (cryoturbation). Sauf dans les régions boréales, les formations de buissons nains situées au-dessus de la limite des arbres en montagne ne devraient pas être appelées toundra, parce qu'elles sont en général plus riches en buissons nains et en herbes, et qu'elles atteignent des tailles supérieures dues au fait que les radiations sont plus importantes dans les latitudes plus basses.

IV.D.1 TOUNDRA SURTOUT DE BRYOPHYTES. Prédominance de tapis ou de petits coussins de mousses chamæphytiques. Les groupes de buissons nains sont en général dispersés de façon irrégulière et pas très denses. Aspect général plus ou moins vert sombre, vert olive ou brunâtre.

141 IV.D.1a *Toundra de mousses avec buissons nains cespiteux.*

142 IV.D.1b *Toundra de mousses avec buissons nains rampants ou en tapis.*

143 IV.D.2 TOUNDRA SURTOUT DE LICHENS. Tapis où les lichens fruticuleux dominent, donnant ainsi à la formation un aspect gris plus ou moins prononcé. Buissons nains surtout sempervirents, rampants ou pulvinés. Toundra de lichens avec buissons nains.

IV.E *FORMATION DE TOURBIÈRE A MOUSSES AVEC BUISSONS NAINS.* Accumulations oligotrophiques de marécage principalement formées de *Sphagnum* ou d'autres mousses, qui en général couvrent assez bien la surface. Les buissons sont concentrés sur les parties relativement sèches ou bien dispersés de façon lâche. A partir d'un certain développement elles ressemblent aux formations de buissons nains sur sol minéral.

Hémicryptophytes graminoïdes, géophytes à rhizomes et autres formes biologiques herbacées peuvent prédominer localement. Les arbres et arbustes à croissance lente peuvent pousser en tant qu'individus isolés ou en groupes, ou en forêt claire, formations qui sont en bordure de la tourbière ou bien qui peuvent être remplacées par des formations ouvertes en succession cyclique. Les divisions suivantes correspondent à la classification des types de tourbière adoptée en Europe.

IV.E.1 TOURBIÈRE ÉLEVÉE. Par la croissance des sphaignes, elle s'élève au-dessus du niveau général de la nappe d'eau et retient une nappe d'eau suspendue. Non approvisionnée en eau " minérale " (c'est-à-dire en eau qui a touché des sols inorganiques) mais seulement en eau de pluie (tourbière réellement ombrotrophique).

144 IV.E.1a *Tourbière typiquement élevée (subocéanique de basse altitude et submontagnarde).* Les mousses dominent, excepté localement sur des hummocks secs élevés, qui sont dominés par des buissons nains. Arbres rares et, lorsqu'ils sont présents, concentrés sur les flancs marginaux de l'accumulation convexe de la tourbière. Principalement entourée par un marécage très humide mais moins oligotrophique.

145 IV.E.1b *Tourbière élevée de montagne (ou subalpine).* Croissance plus lente que celle de la tourbière élevée typique (ou formée en une période antérieure de climat plus chaud et actuellement stabilisée ou en cours de destruction par l'érosion). Souvent couverte de cypéracées ou de buissons nains sempervirents. Micro- ou nano-phanérophytes (ex.: *Pinus mughus*) localement dominantes.

146 IV.E.1c *Tourbière subcontinentale de forêt claire.* Temporairement couverte de formations arborées ouvertes de faible productivité, qui peuvent être remplacées par des formations de *Sphagnum* comparables à IV.E.1a, dans les années les plus humides.

IV.E.2 TOURBIÈRE NON ÉLEVÉE. Peu ou pas élevée de façon marquée au-dessus de la nappe minéralisée du paysage environnant. Donc en général plus humide et pas aussi oligotrophique que IV.E.1. Plus pauvre en mousses que la tourbière élevée typique (IV.E.1a), vers laquelle plusieurs formes de transition sont possibles.

147 IV.E.2a *Tourbière en tapis (basses altitudes océaniques, submontagnarde ou montagnarde).* La micro-surface de la tourbière est moins ondulée et moins riche en mousses de croissance active que IV.E.1a. Buissons sempervirents épars de même que les hémicropyptophytes cespiteuses (cypéracées ou graminées) et quelques géophytes rhizomateuses.

148 IV.E.2b *Tourbière en bourrelets (tourbière finnoise " Aapa ").* Tourbière plane oligotrophique, avec hummocks en bourrelets alignés dans les régions basses boréales. L'adjectif finnois indique une tourbière ouverte sans arbres ou avec seulement quelques arbres très peu vigoureux, qui poussent sur d'étroits bourrelets bas allongés. Ces bourrelets de tourbière sont formés par la pression de la glace qui couvre les parties de tourbière plus ou moins inondées, depuis les premières précipitations jusqu'à la fin du printemps. Seuls ces bourrelets sont couverts de buissons nains et sont riches en sphaignes. La majeure partie de la tourbière est comparable à un marécage humide à *Carex*.

Végétation herbacée

La classification de la végétation herbacée mérite un examen approfondi, pour les raisons principales suivantes: *a*) les changements saisonniers continuels de la physionomie des communautés; *b*) les problèmes que pose la distinction entre nombre de prairies tropicales et de prairies non tropicales; *c*) l'exploitation des prairies, qui peut influer profondément sur la structure de la végétation et qui peut changer fréquemment; *d*) les problèmes que pose la distinction entre les prairies naturelles et les prairies artificielles.

Il existe deux grands types de végétation herbacée: les graminoïdes et les forbes. Les graminoïdes comprennent toutes les graminées herbacées et autres plantes d'aspect herbacé comme les laiches (*Carex*), les joncs (*Juncus*), les massettes (*Typha*), etc. Les forbes sont des plantes herbacées à feuilles larges, comme le trèfle (*Trifolium*), les hélianthes (*Helianthus*), les fougères, les gentianes (*Asclepias*), etc. D'ordinaire, toutes les plantes herbacées non graminoïdes sont comprises dans les forbes.

La classification Unesco emploie fréquemment le terme de « prairie » pour désigner un type de végétation herbacée où dominent les formes graminoïdes. Comme les graminoïdes comprennent de nombreux taxa autres que les graminées, le mot « prairie » doit être compris ici comme désignant un type physionomique de végétation sans portée floristique.

Comme dans le cas des communautés ligneuses, la hauteur est une donnée essentielle qui caractérise une communauté végétale dominée par des formes herbacées. La hauteur des plantes herbacées étant sujette à de grandes variations saisonnières, il faut la mesurer (ou l'évaluer) au moment de la floraison, c'est-à-dire lorsque les inflorescences ont atteint tout leur développement. Il est possible que les inflorescences n'apparaissent pas lorsque la prairie est pâturée régulièrement ou intensément; leur hauteur doit alors être évaluée.

La couverture est une autre caractéristique importante de la végétation herbacée. Mais, à l'échelle de 1/1 000 000 ou moins, toutes les communautés herbacées sont présumées avoir une couverture plus ou moins continue, si bien que ce facteur n'apparaît pas dans la légende des cartes. La très faible densité constitue une exception. Dans ce cas, la communauté est dite ouverte.

Les formes de développement des graminoïdes méritent d'être notées, et l'on distingue ordinairement la forme en motte et la forme en touffe ou en hummock ou des combinaisons de ces deux formes. Dans la classification de l'Unesco, toutes les communautés de graminoïdes sont considérées comme ayant plus ou moins la forme motteuse; aussi n'est-il pas nécessaire de fournir cette précision. Mais, lorsque les graminées en touffe dominent, cette particularité influe profondément sur la physionomie de la végétation. La forme de graminées en touffe doit donc figurer dans la description des communautés végétales intéressées. Ces descriptions font pendant aux descriptions des formes des houppiers d'arbres sempervirents à aiguilles, comme par exemple aux alinéas I.A.9c et d.

Les types de végétation herbacée comprennent souvent une synusie de plantes ligneuses qui donnent au type un caractère particulier. Souvent donc ces synusies servent à caractériser les communautés herbacées. Parmi les particularités relativement importantes d'une synusie ligneuse figurent la hauteur et la densité, qu'il s'agisse de végétaux sempervirents ou décidus, à aiguilles, à feuilles larges ou aphylles (essentiellement dépourvus de feuilles), etc. Ces particularités dominent dans des combinaisons très diverses et sont très évocatrices de la physionomie générale de la végétation. D'autres particularités peuvent être admises si elles caractérisent une communauté assez étendue pour figurer sur une carte à l'échelle de 1/1 000 000 ou moins, par exemple la nature sclérophylle de la synusie ligneuse de nombreuses combinaisons *Eucalyptus*-graminées en Australie.

De nombreux termes, fréquemment employés, servent à désigner ces formes de végétation: savane, steppe, pelouse, etc. On les a évités dans les définitions de la classification Unesco parce qu'ils donnent lieu à trop d'interprétations contradictoires. Parfois, cependant, on les a ajoutés entre parenthèses lorsque cela aide le lecteur à identifier la catégorie. Il vaut mieux toujours employer la terminologie de l'Unesco sur les cartes de végétation. Des termes d'emploi local courant, significatifs pour les habitants des régions considérées (par exemple "campo cerrado") peuvent être ajoutés; mais ils ne sauraient remplacer les termes de l'Unesco dans les légendes des cartes. Ainsi les cartes de végétation ont-elles un sens aussi bien pour les utilisateurs locaux que pour un public mondial. Cela revêt une importance particulière pour les études comparatives.

V VÉGÉTATION HERBACÉE

V.A *VÉGÉTATION GRAMINOÏDE HAUTE.* Dans les prairies hautes, les formes graminoïdes dominantes ont plus de 2 mètres de hauteur lorsque les inflorescences sont en plein épanouissement. Des forbes peuvent être présentes; mais leur taux de couverture est inférieur à 50%.

V.A.1 PRAIRIE HAUTE AVEC UNE SYNUSIE ARBORÉE COUVRANT DE 10 A 40%, avec ou sans buissons. Cela ressemble quelque peu à une forêt très claire, avec une couverture de sol plus ou moins continue (plus de 50%) de hautes graminoïdes. Pour les catégories ayant une synusie arborée couvrant plus de 40%, voir la classe de formations II.

149 V.A.1a *Synusie arborée de sempervirents latifoliés.*

150 V.A.1b *Synusie arborée de semi-sempervirents latifoliés,* c'est-à-dire composée de 25% au moins d'arbres sempervirents latifoliés et d'un pourcentage analogue d'arbres décidus latifoliés.

V.A.1c *Synusie ligneuse de décidus latifoliés.*
151 V.A.1c(1) Analogue à la précédente, mais inondée saisonnièrement (ex.: en Bolivie du Nord-Est).

V.A.2 PRAIRIE HAUTE AVEC UNE SYNUSIE ARBORÉE COUVRANT MOINS DE 10%, avec ou sans buissons. Subdivisions V.A.2a à c comme dans V.A.1.

152 V.A.2a

153 V.A.2b

154 V.A.2c

155 V.A.2d *Prairie tropicale ou subtropicale haute avec arbres et (ou) buissons poussant en touffes sur des termitières* (savane à termites).

V.A.3 PRAIRIE HAUTE AVEC UNE SYNUSIE DE BUISSONS (savane à buissons).

156 V.A.3a (Subsivisions comme dans V.A.2.)

157 V.A.3b

158 V.A.3c

159 V.A.3d

V.A.4 PRAIRIE HAUTE AVEC UNE SYNUSIE LIGNEUSE COMPOSÉE SURTOUT DE PLANTES A TOUFFE (ordinairement palmiers).

160 V.A.4a *Prairie tropicale à palmiers* (ex.: la savane à palmiers d'*Arocomia totai* et *Attalea princeps* du nord de Santa Cruz de la Sierra, en Bolivie).
161 V.A.4a(1) Comme ci-dessus, inondée saisonnièrement (ex.: la savane à *Mauritia vinifera,* dans les llanos de Mojos, en Bolivie).

V.A.5 PRAIRIE HAUTE A PEU PRÈS DÉPOURVUE DE SYNUSIE LIGNEUSE.

162 V.A.5a *Prairie tropicale* telle que décrite en V.A.5 comme dans diverses régions des basses latitudes de l'Afrique.
163 V.A.5a(1) Comme ci-dessus, inondée saisonnièrement (ex.: les campos de Várzea de la basse vallée de l'Amazone).
164 V.A.5a(2) Comme ci-dessus, mouillée ou inondée la plus grande partie de l'année (ex.: le marécage à papyrus (*Cyperus papyrus*) de la haute vallée du Nil).

V.B — *PRAIRIE MOYENNE.* Les formes graminoïdes dominantes ont de 50 cm à 2 mètres de hauteur lorsque leurs inflorescences sont en plein épanouissement. Des forbes peuvent être présentes, mais elles couvrent moins de 50%.

165 V.B.1a — (Divisions et subdivisions comme dans V.A.1 à V.A.3d.)

166 V.B.1b

167 V.B.1c

168 V.B.2a

169 V.B.2b

170 V.B.2c

171 V.B.2d

172 V.B.3a

173 V.B.3b

174 V.B.3c

175 V.B.3d

176 V.B.3e — *Synusie ligneuse se composant principalement de buissons épineux décidus* (ex.: la savane tropicale à broussailles épineuses de la région sahélienne d'Afrique, à *Acacia tortilis*, *A. senegal*, etc.).

V.B.4 — PRAIRIE MOYENNE AVEC UNE SYNUSIE OUVERTE DE PLANTES A TOUFFE, généralement palmiers.

177 V.B.4a — *Prairie subtropicale moyenne avec bosquets clairs de palmiers* (ex.: à Corrientes, en Argentine).

178 V.B.4a(1) — Comme ci-dessus, inondée saisonnièrement (ex.: bosquet à palmiers *Mauritia* dans les llanos de Colombie et du Venezuela).

V.B.5 — PRAIRIE MOYENNE A PEU PRÈS DÉPOURVUE DE SYNUSIE.

179 V.B.5a — *Prairie moyenne composée principalement de graminées en mottes* (ex.: la prairie à hautes graminées du Kansas oriental).

180 V.B.5a(1) — Mouillée ou inondée la plus grande partie de l'année (ex.: les marécages à *Typha*).

181 V.B.5a(2) — Sur sol sableux ou dunes (ex.: les communautés d'*Andropogon hallii* dans les dunes de sable du Nebraska).

182 V.B.5b — *Prairie moyenne composée surtout de graminées cespiteuses* (ex.: les prairies à touffes dures [*Festuca novaezelandiae*] de Nouvelle-Zélande).

V.C — *PRAIRIE BASSE.* Les formes graminoïdes dominantes ont moins de 50 cm de hauteur lorsque leurs inflorescences sont épanouies. Des forbes peuvent être présentes, mais elles couvrent moins de 50%.

183 V.C.1a — Divisions et subdivisions comme de V.B.1 à V.B.4.)

184 V.C.1b

185 V.C.1c

186	V.C.2a	
187	V.C.2b	
188	V.C.2c	
189	V.C.2d	
190	V.C.3a	
191	V.C.3b	
192	V.C.3c	
193	V.C.3d	
194	V.C.3e	
195	V.C.4a	
196	V.C.4a(1)	
197	V.C.5a	*Communauté herbacée cespiteuse alpine tropicale, ouverte à fermée, avec une synusie ligneuse de plantes à touffe (Espeletia, Lobelia, Senecio)*, de fruticées naines microphylles à leptophylles et de plantes pulvinées, souvent à feuilles cotonneuses. Au-dessus de la limite des arbres aux basses latitudes: Páramo et types de végétation apparentés sans neige dans les régions alpines du Kenya, de Colombie, du Venezuela, etc.
198	V.C.5b	Analogues à V.C.5a, mais très ouvertes et sans plantes à touffe; chutes de neige nocturnes fréquentes (neige disparue dès 9 heures du matin). Le super-Páramo (c'est-à-dire au-dessus du Páramo) de J. Cuatrecasas [1].
199	V.C.5c	*Végétation herbacée cespiteuse alpine des régions tropicales ou subtropicales* avec peuplements clairs de sempervirents avec ou sans buissons nains ou buissons décidus (ex.: la puna).
	V.C.5c(1)	Avec nombreuses formes succulentes.
200	V.C.5d	*Végétation herbacée cespiteuse de diverses couvertures avec buissons nains.*
	V.C.5d(1)	Avec plantes en coussin qui peuvent être localement plus importantes que les buissons nains (ex.: la puna du sud d'Oruro, en Bolivie).
	V.C.6	PRAIRIE BASSE A PEU PRÈS DÉPOURVUE DE SYNUSIE LIGNEUSE.
201	V.C.6a	*Communauté herbacée basse.* Sa structure et sa composition floristique peuvent varier en fonction des précipitations, elles-mêmes très variables, d'un climat semi-aride (ex.: la prairie à herbes courtes [*Bouteloua gracilis* et *Buchloë dactyloïdes*] de l'est du Colorado).
202	V.C.6b	*Communauté herbacée cespiteuse* (ex.: les communautés de canche bleue ou " blue tussock " [*Poa colensoi*] de Nouvelle-Zélande et la puna alpine sèche à *Festuca orthophylla* du nord du Chili et du sud de la Bolivie).
	V.C.7	PRAIRIE (PELOUSE) MOYENNE A BASSE, A MÉSOPHYTES.
203	V.C.7a	*Communauté à gazon formant mottes*, souvent riche en forbes, généralement dominée par des hémicryptophytes. Se trouve surtout à basse altitude sous climat frais et humide en Amérique du Nord et en Eurasie. De nombreuses plantes peuvent rester au moins partiellement vertes pendant l'hiver, même sous la neige aux hautes latitudes.
204	V.C.7b	*Pelouse alpine et subalpine des hautes latitudes*, par opposition aux types de végétation du Páramo et de la puna aux basses latitudes. Ordinairement humide une grande partie de l'été grâce à l'eau de fusion.

1. "Páramo vegetation and its life forms", *Colloquium geographicum*, vol. 9, Bonn, F. Dummlers Verlag, 1968, 223 p.

205 V.C.7b(1) Riche en forbes, par exemple sur l'Olympic Peninsula (Wash.).

206 V.C.7b(2) Riche en buissons nains, par exemple sur les montagnes Rocheuses du Colorado.

207 V.C.7b(3) Communauté en plaques nivales. Communauté ouverte, riche en petites forbes et (ou) en buissons nains comparables à des forbes (ex.: *Salix herbacea*). L'équivalent aux hautes latitudes du super-Páramo des basses latitudes (voir V.C.5b).

208 V.C.7b(4) Prairie des couloirs d'avalanche, se présentant en bandes étroites de prairie entre des forêts sur les versants escarpés des hautes montagnes où les avalanches, qui descendent chaque année au printemps, empêchent la forêt de s'établir. Structure variable; il peut y avoir quelques buissons ou arbres endommagés.

V.C.8 TOUNDRA DE GRAMINOÏDES. Comme dans le cas de la toundra de buissons nains (sous-classe de formations IV.D), l'emploi du terme " toundra de graminoïdes " doit être limité aux hautes latitudes, c'est-à-dire à celles qui sont au-delà de la limite polaire des arbres.

209 V.C.8a *Toundra de graminoïdes à forme cespiteuse.* La plupart des graminoïdes poussent en hammocks. Entre les hammocks croissent souvent des mousses et (ou) des lichens (ex.: la toundra à *Eriophorum* du nord de l'Alaska).

210 V.C.8a(1) Inondée saisonnièrement.

211 V.C.8b *Toundra de graminoïdes en mottes.* Les graminoïdes forment une motte plus ou moins dense, avec souvent diverses forbes et divers buissons nains (cespiteux et rampants). Les lichens peuvent être fréquents (ex.: la toundra près du lac d'Iliama, en Alaska).

212 V.C.8b(1) Inondée saisonnièrement.

V.D *VÉGÉTATION A FORBES.* Les communautés végétales se composent principalement de forbes, c'est-à-dire que les forbes couvrent plus de 50%. Des graminoïdes peuvent être présentes, mais elles couvrent moins de 50%, souvent beaucoup moins.

V.D.1 COMMUNAUTÉ DE FORBES HAUTES. Les forbes dominantes ont plus d'un mètre de hauteur lorsqu'elles atteignent leur plein développement.

213 V.D.1a *Surtout forbes pérennes à fleurs, et fougères.*

214 V.D.1a(1) Substrat salin et (ou) mouillé une grande partie de l'année (ex.: les marécages à *Batis-Salicornia* de Floride et les formations à *Heliconia-Galathea* d'Amérique centrale).

215 V.D.1b *Fougeraies*, parfois en peuplements presque purs, notamment en climat humide (ex.: *Pteridium aquilinum*).

216 V.D.1c *Surtout forbes annuelles.*

V.D.2 COMMUNAUTÉ DE FORBES BASSES, dominée par des forbes qui n'atteignent pas un mètre à leur plein développement.

217 V.D.2a *Surtout forbes pérennes à fleurs, et fougères* (ex.: pelouse à *Celmisia* de Nouvelle-Zélande et pelouse à forbes des Aléoutiennes, en Alaska).

218 V.D.2b *Surtout forbes annuelles.*

219 V.D.2b(1) Communauté éphémère de forbes dans les régions tropicales et subtropicales où les précipitations sont très faibles et où, de l'automne au printemps, les nuages humectent la végétation et le sol (ex.: sur les collines côtières du Pérou et du nord du Chili). Dominée par des forbes annuelles qui germent au début de la saison nuageuse et poussent abondamment jusqu'à la fin de celle-ci, en donnant au paysage un aspect frais et vert. Pendant la saison sèche, l'aspect est désertique. Géophytes et hémicryptophytes cryptogamiques ou chamæphytes sont constamment présentes et peuvent devenir localement dominantes. Éventuellement quelques phanérophytes, reliques d'une forêt ombrophile naturelle.

220 V.D.2b(2) Communauté éphémère ou épisodique de forbes des régions arides: le " désert fleuri ". Surtout forbes à croissance rapide, parfois concentrées dans les dépressions où l'eau peut s'accumuler. Parfois cette communauté peut former des synusies en formations buissonneuses ou buissonneuses naines des régions arides [voir IV.C]. (Ex.: dans le désert Sonorien.)

221 V.D.2b(3) Communauté épisodique de forbes. Communauté irrégulière de structure et de composition variables se développant sur les parties sèches des lits de rivière lorsque la période des basses eaux dure plus de deux mois, ou sur les zones temporairement asséchées des lits de rivière lorsque le chenal a tendance à changer plus ou moins fréquemment (ex.: les lits des grands cours d'eau des llanos de la Colombie et du Venezuela).

V.E *VÉGÉTATION HYDROMORPHE DES EAUX DOUCES* (végétation aquatique). Comme dans le cas des communautés aquatiques de mangroves, les communautés d'hélophytes ne sont pas classées dans cette catégorie. Ainsi, les peuplements de *Typha* sont considérés comme des communautés de graminoïdes moyennes, mouillées ou submergées la plus grande partie de l'année [cf. V.B.5a(1)].

V.E.1 COMMUNAUTÉ ENRACINÉE DES EAUX DOUCES. Composée de plantes aquatiques que l'eau supporte structurellement, c'est-à-dire qui n'ont pas de support propre, à la différence des hélophytes.

222 V.E.1a *Formation tropicale et subtropicale de forbes*, sans contrastes saisonniers appréciables (ex.: les communautés de *Victoria regia* de l'Amazone).

223 V.E.1b *Formation de forbes des moyennes et hautes latitudes* à contrastes saisonniers importants (ex.: les communautés de *Nymphae* et de *Nuphar*).

V.E.2 COMMUNAUTÉ FLOTTANTE LIBRE DES EAUX DOUCES.

224 V.E.2a *Formation flottante libre des régions tropicales et subtropicales* (ex.: *Pistia*, *Eichhornia*, *Azolla pinnata*, etc.).

225 V.E.2b *Formation flottante libre des latitudes moyennes et hautes*, disparaissant l'hiver (ex.: *Utricularia purpurea*, *Lemna*).

Cartographie

LES LIMITES

Les diverses unités distinguées ci-dessus doivent être limitées sur la carte par un trait qui les sépare des unités voisines. Dans une carte géologique, la chose est facile. Pour les cartes climatiques ou de végétation, la représentation est souvent bien plus difficile, car les unités à représenter ont rarement des limites très nettes. C'est vrai pour une forêt, pour un marécage, mais souvent il y a transition continue d'une unité à la voisine. Typographiquement il est très difficile de réaliser des dégradés et bien des limites donnent une précision qui n'existe pas dans la nature.

Pour une échelle de 1/5 000 000, les contours donnés par les cartes déjà publiées sont parfois insuffisamment précis. L'utilisation de photographies aériennes, surtout prises à haute altitude, peut permettre de préciser les contours de formations dont on connaît les caractères botaniques.

Pour les échelles plus grandes, l'emploi des photographies aériennes est nécessaire. Il faut naturellement que le fond topographique soit exact, ce qui n'est pas partout réalisé.

COULEURS A EMPLOYER

Suivant les idées appliquées depuis 1924 par H. Gaussen, les principes sont théoriquement les suivants:

1. La façon de mettre la couleur — teinte plate, grisé de lignes, grisé de points — représente les types physionomiques essentiels: forêt, broussaille, herbacées.
2. La couleur employée indique les conditions de l'environnement par synthèse de couleurs élémentaires représentant chacune un facteur du milieu. Cette synthèse, réalisée en 1926, a donné par la suite de bons résultats dans les cartes publiées au 1/5 000 000 (Unesco-FAO: pays méditerranéens), au 1/1 000 000 (Inde, Sahara, Madagascar), 1/200 000 (végétation de la France et de l'Afrique du Nord).

Quand on a pris l'habitude du choix des couleurs, du premier coup d'œil on obtient une vision des conditions générales de la végétation. Ces conditions sont essentiellement climatiques aux petites échelles (1/5 000 000) mais l'importance de certains sols peut apparaître (dunes, mangroves, marécages, lacs).

On trouvera plus loin un dépliant où sont donnés des exemples des principales couleurs.

Dans les pays tropicaux, où les conditions de l'environnement sont très différentes en saison sèche et en saison humide, on utilise des bandes verticales alternantes, représentant les deux types.

Pour chaque couleur employée, 3 intensités sont utilisées; par exemple, pour la couleur bleue (B): B_1, très léger; B_2, grisé assez vigoureux; B_3, teinte plate.

Les couleurs fondamentales sont: rouge = R; orange = O; jaune = Y (*yellow* en englais); vert = V; bleu = B; violet = P (pourpre violet); gris = G; marron = M (pour les tourbières).

Des superpositions permettent de créer de nombreuses nuances.

La couleur écologique indiquée par les numéros des formations (chiffres gras, colonne de gauche de la Classification) résulte de la superposition de couleurs élémentaires représentant les divers facteurs du milieu. A l'échelle de 1/5 000 000 les facteurs essentiels sont l'humidité et la chaleur.

Les gammes sont les suivantes dans l'ordre des couleurs de l'arc-en-ciel.

Facteur humidité

Très sec: orange
Humidité moyenne: jaune
Très humide: bleu
avec tous les intermédiaires.

Facteur chaleur

Très chaud: rouge
Chaleur moyenne: jaune
Très froid: gris
avec tous les intermédiaires.

Une forêt tempérée, donc un peu humide et de température moyenne, est
bleu clair + jaune = vert clair.

Une forêt tropicale, donc humide et chaude, est
bleu + rouge = violet.

Un désert chaud, donc sec et chaud, est
orange + rouge = rouge orangé.

Un désert froid, donc sec et froid, est
orange + gris = gris orangé.

Etc.

L'absence de végétation se signale de la manière suivante:

blanc: avec chaud = petits points espacés rouges,
très froid = petits points espacés gris.

La nature du sol intervient parfois: sable, marais, mangrove, etc. Des signes spéciaux peuvent la représenter.

SIGNES EMPLOYÉS

Les signes sont inspirés de ceux qui ont été publiés par H. Gaussen et acceptés par Braun-Blanquet. Ces signes sont présentés pour les cartes à grande échelle. Chaque genre a un type de signe et les diverses espèces sont des variations autour de ce type.

Aux petites échelles, dans une carte physionomique, on ne peut pas entrer dans autant de détails.

On a distingué les formes: arbre feuillu, de grande taille, de taille moyenne, buissonnant. De même pour les résineux.

On a utilisé des signes complémentaires pour: épineux, succulents, microphylles, tourbières, etc.

Les plantes à feuilles caduques sont dessinées sans plage noire.

Les plantes à feuilles persistantes sont dessinées en noir.

Les savanes, les pelouses ont des signes spéciaux.

Les marécages, l'eau douce, l'eau salée continentale ont aussi leurs symboles.

Annexe pour la représentation des cultures

En matières cartographique et géographique en général, la question d'échelle est fondamentale.

Alors qu'à l'échelle de 1/5 000 000 il n'est pas possible de représenter les cultures malgré leur grand intérêt économique, à l'échelle de 1/1 000 000 et à fortiori aux échelles plus grandes, on peut inscrire sur la carte l'essentiel des statistiques agricoles sur un fond blanc. La végétation potentielle des parties actuellement cultivées, souvent difficile à connaître, est fournie par un carton annexé à la carte et à couleurs écologiques. Les limites des diverses unités qu'on y distingue sont indiquées sur la carte principale.

Pour ces échelles, la carte est à la fois botanique et agricole, ce qui augmente beaucoup son intérêt pour le public non spécialiste en botanique. On en connaît divers exemples, tels que la carte de l'Inde au 1/1 000 000 ou celle de la France au 1/200 000.

Quand l'échelle permet de représenter les cultures, chaque signe a une valeur statistique. Par exemple, un signe de vigne représente 10 ha, un signe d'olivier 10 000 ou 1 000 ou 100 arbres, suivant sa grosseur, etc.

L'emploi de lettres permet d'indiquer le pourcentage du type d'utilisation du sol dans une division administrative déterminée (par exemple, canton). Tout cela dépend de l'échelle mais est utilisable au 1/1 000 000 et aux échelles plus grandes.

L'avantage du fond blanc est qu'on peut indiquer des signes en couleur en rapport avec l'écologie de la plante représentée.

Symboles de végétation

	Sempervirents	Caducifoliés
ARBRES		
Feuillus		
Résineux		
De grande taille > 50 mètres		
Conifères genre pins		
Bouteilles		
Sur termitières		
Microphylles		
Épineux		
Palmiers		
ARBUSTES		
Feuillus		
Grandes fruticées		
Fruticées naines, sèches, éventuellement décidues		
Sur termitières		
Microphylles		
Succulents		
Bambous		
Buissons		
Palmiers nains		
Fruticées naines		
Fruticées naines en coussins		
HERBACÉES		
Forbes pérennes		
Forbes éphémères, épisodiques		
Fougères		
Lichen		
Mousse		
Salicornes		
Rampants		
Végétations aquatique flottante, enracinée		
Marécages		
Tourbières		
Dunes		

Clasificación internacional y cartografía de la vegetación

Índice

Introducción

La siguiente clasificación ha sido preparada por el Comité Permanente de la Unesco para la Clasificación y Cartografía de la Vegetación sobre una Base Mundial. El objetivo es ofrecer un esquema amplio de las categorías más importantes que se usarán en mapas de vegetación de escalas 1/1 000 000 o menores [1].

Aunque la clasificación esté dirigida principalmente hacia la preparación de nuevos mapas, también puede servir para transponer a este sistema los mapas de vegetación existentes.

El Comité revisó las clasificaciones existentes y concluyó que ninguna de ellas era enteramente apropiada para el objetivo propuesto; a pesar de esto las clasificaciones existentes han tenido su influencia en la opinión y las decisiones del comité. Como una clasificación que es, el sistema es necesariamente artificial, las relaciones ecológicas no están contenidas en el ordenamiento de las unidades.

Las unidades en esta clasificación son unidades de vegetación e incluyen formaciones zonales y las azonales de más importancia y distribución, así como formaciones intervenidas. Es difícil usar el enfoque florístico como una base general de clasificación mundial, porque formaciones equivalentes en diferentes partes del mundo están constituidas de floras diferentes, sin embargo, en un área particular la florística juega generalmente un papel significativo en la definición y distinción de las comunidades.

1. El sistema puede ser también a escalas mayores extendiéndolo a subdivisiones menores. Se presentan unas pocas subdivisiones menores que dirven como ejemplos.

El mejor conjunto de variables para una comparación a nivel mundial son la fisionomía y la estructura de la vegetación. Desgraciadamente, estos dos atributos de la vegetación no son siempre claramente identificables en relación con habitats o medios ecológicos. Por esta razón los términos que se refieren al clima, suelo y formas del terreno han sido incluidos en los nombres y ocasionalmente en las definiciones, en donde pueden ayudar a identificar una unidad. En la preparación de la clasificación, por lo tanto, se estimó conveniente ser pragmáticos en vez de estrictamente sistemáticos, para proveer unidades con nombres y definiciones que fueran cortas, significativas y descriptivas.

Aunque la mayoría de las unidades se definen fisionómicamente, todas las unidades indican condiciones ambientales. La fisionomía de muchas formaciones graminoides no permite a un cartógrafo reconocer sus afinidades latitudinales. Por consiguiente, indicar esas afinidades en la clase de formaciones V de esta clasificación fisionómica conduciría a un número innecesario de duplicaciones y repeticiones. Sin embargo, se invita a los autores a emplear términos tales como tropical, templado, etc., en relación con las comunidades de plantas herbáceas cuando preparen sus leyendas de mapas y siempre que esa información suplementaria ayude al lector del mapa de vegetación a identificar las categorías de que se trate.

Uno de los problemas que se ha planteado con más insistencia en relación con la clasificación de la Unesco se refiere a los aspectos dinámicos de la vegetación y al lugar que ocupan las diferentes categorías en la sucesión hacia la climax. Evidentemente, no todas las categorías representan condiciones climácicas. En este caso, lo mismo que en el empleo de información mesológica suplementaria, el Comité se inspiró en principios pragmáticos. A la escala de 1/1 000 000 o menor, no es factible indicar los detalles que con frecuencia se requieren para identificar las comunidades climácicas. Especialmente, en lo que respecta a las comunidades herbáceas, la interpretación de la vegetación en términos de climax es a menudo muy difícil o imposible. En consecuencia, la clasificación de la Unesco se basa en los tipos de vegetación climácica siempre que es practicable. Además, también se incluyen las condiciones cercanas a la climax y los tipos de vegetación «seminatural» cuando existen en el momento de la observación. Esto ofrece al cartógrafo mayor flexibilidad y le permite representar muchos tipos de vegetación cuya situación sucesional es incierta, se presta a diversas interpretaciones o es imposible determinar actualmente. Se excluye la vegetación cultivada en sentido estricto, es decir, la denominada vegetación mesicol [1]. Que los usuarios de los mapas de vegetación basados en la clasificación de la Unesco no esperen encontrar campos de trigo, viñedos, plantaciones de plátanos ni arrozales. Esas indicaciones no tienen cabida en esta clasificación [2].

Por lo tanto, puede decirse que la clasificación de la Unesco tiene un carácter fundamentalmente fisionómico estructural con información ecológica suplementaria integrada en sus diferentes categorías y aplicable a la vegetación natural y seminatural. Aunque relativamente completa, está organizada de forma que permita la adición de nuevas categorías cuando sea conveniente.

1. A. W. Küchler, "Natural and cultural vegetation", *Professional geographer*, vol. 21, p. 383-385.
2. Es evidente que sería difícil indicar la climax en las partes cultivadas, ya que a menudo sería, por ejemplo un bosque, pero los cultivos ocupan un lugar considerable en la superficie de la tierra y hay que tenerlos muy en cuenta en la cartografía. Ateniéndose a las características fisionómicas los cereales serán graminoides, la remolacha y la patata serán forbias y la viña y los árboles frutales serán plantas arbustivas. En el anexo, p. 92, se indicará el método utilizado para la representación gráfica de los cultivos. (Nota de H. Gaussen.)

Clasificación

En la clasificación de la Unesco, las unidades de diferente rango o jerarquía están representadas de la forma siguiente:

I, II, etc. = CLASES DE FORMACIÓN
A, B, etc. = *SUBCLASES DE FORMACIÓN*
1, 2, etc. = GRUPOS DE FORMACIÓN
a, b, etc. = *Formación*
(1), (2), etc. = Subformación
(a), (b), etc. = Otras subdivisiones

I — BOSQUE DENSO
Formado por árboles de más de 5 metros de altura, cuyas copas se tocan [1].

I.A — *BOSQUE MAYORMENTE SEMPERVIRENTE.* El dosel superior nunca está sin follaje, sin embargo, algunos árboles individualmente pueden perder sus hojas.

I.A.1 — BOSQUE TROPICAL OMBRÓFILO. (Convencionalmente llamado " bosque pluvial tropical ".) Formado principalmente por árboles sempervirentes, generalmente con yemas desnudas —presentándose pocos casos de yemas protegidas, sin resistencia al frío y a la sequía. Sempervirentes verdaderos, es decir, el dosel superior del bosque permanece verde a lo largo del año, pero algunos árboles individualmente pueden permanecer desnudos durante algunas semanas solamente y no simultáneamente con todos los demás. Las hojas de muchas especies son acuminadas.

1 I.A.1a — *Bosque tropical ombrófilo de baja altitud.* Compuesto generalmente de numerosas especies de crecimiento rápido, muchas de ellas alcanzando más de 50 metros de altura [2], generalmente con corteza lisa; a menudo gruesa, algunas con aletones (raíces tabulares). A menudo se presentan árboles emergentes o por lo menos el dosel superior es irregular en altura. Sotobosque poco denso, y compuesto de la regeneración. Palmeras

1. En estado de reproducción o en estado de crecimiento secundario inmaduro pueden tener menos de 5 metros de altura temporalmente, pero formados por individuos con solamente un tronco (es decir, árboles verdaderos y no arbustos). En condiciones subpolares el límite puede ser de sólo 3 metros y en condiciones tropicales de 2 a 10 metros.
2. Los límites de altura son sólo una guía generalizada y no un criterio absoluto.

y otros árboles estipitados raros, lianas casi ausentes excepto seudo-lianas (es decir, plantas que se originan en las ramas de los árboles y posteriormente enraizan en el suelo o viceversa). Líquenes costrosos y algas verdes son las únicas formas de vida epífita presentes constantemente; las epífitas vasculares generalmente no son abundantes; abundantes solamente en situaciones excesivamente húmedas (p.e. Sumatra, Atrato Valley [Columbia] etc.)

2 I.A.1b — *Bosque tropical ombrófilo submontano.* Árboles emergentes casi siempre ausentes y el dosel superior regular en altura. En el sotobosque son comunes las latifoliadas herbáceas. Las epífitas vasculares y las seudo-lianas son abundantes (p.e. vertiente atlántica de Costa Rica).

I.A.1c — *Bosque tropical ombrófilo montano.* (Corresponde más exactamente al bosque pluvial tropical virgen —Virgin Tropical Rainforest— que aparece en los textos.) Epífitas vasculares y de otra clase abundantes. La altura de los árboles es de menos de 50 metros; las copas se extienden relativamente hacia abajo del tallo. La corteza a menudo es más o menos rugosa. El sotobosque abundante, a menudo representado por macro- y micro-fanerófitas arosetadas (helechos arborescentes o palmeras pequeñas). El piso es rico en hierbas higromorfas y criptógamas (p.e. Sierra de Talamanca, Costa Rica).

3 I.A.1c(1) — Latifoliada, la forma más común.

4 I.A.1c(2) — Aciculifoliada.

5 I.A.1c(3) — Microfoliada.

6 I.A.1c(4) — Bambú, rico en graminoides arborescentes (tree-grasses) que reemplazan ampliamente a las macro- y micro-fanerófitas estipitadas.

7 I.A.1d — *Bosque tropical ombrófilo subalpino.* (No incluye bosques o matorrales nublados, esta formación es considerada por algunos investigadores pero probablemente no es importante. Es necesario definirla.)

I.A.1e — *Bosque tropical ombrófilo nublado.* Las copas, ramas y troncos, así como las lianas recargadas de epífitas, principalmente biófitas camefíticas. También el piso está cubierto de camefíticas higromórficas (p.e. *Selaginella* y helechos herbáceos). Los árboles son a menudo retorcidos, de corteza rugosa y rara vez exceden los 20 metros de altura (p.e. Blue Mountains, Jamaica).

8 I.A.1e(1) — Latifoliadas, la forma más común.

9 I.A.1e(2) — Aciculifoliadas.

10 I.A.1e(3) — Microfoliadas.

I.A.1f — *Bosque tropical ombrófilo aluvial.* Similar al bosque sub-montano I.A.1b, pero más rico en palmeras y en formas de vida del sotobosque, particularmente en latifoliadas herbáceas altas (p.e. Musaceae); aletones (raíces tabulares) frecuentes (p.e. cuenca amazónica).

11 I.A.1f (1) — Ripícolas o de galería (en los bancos bajos frecuentemente inundados de los cursos de agua), mayormente dominados por árboles de crecimiento rápido, sotobosque herbáceo casi ausente, epífitas muy raras, pobre en especies (p.e. igapó amazónico).

12 I.A.1f (2) — Ocasionalmente inundado sobre terrazas relativamente secas que se presentan en cursos de agua permanentes. Más epífitas que en las subformaciones (1) y (3), muchas lianas.

13 I.A.1f (3) — Estacionalmente anegados (a lo largo de cursos en los cuales el agua se acumula en grandes áreas planas por varios meses, especialmente detrás de diques naturales bajos); los árboles tienen frecuentemente zancos (raíces fúlcreas); la densidad del dosel superior no es uniforme; por regla general con sotobosque pobre excepto en los lugares más abiertos (p.e. várzea amazónica).

I.A.1g — *Bosque tropical ombrófilo pantanoso.* (No a lo largo de los cursos de agua, pero en habitats más húmedos edáficamente, los cuales pueden ser suplidos de aguas dulces o salobres.) Similar al bosque aluvial I.A.1f, pero por regla general más pobre en especies arbóreas;

muchos árboles tienen aletones (raíces tabulares) o neumatóforos; en su mayoría de más de 20 metros de altura (p.e. Sumatra oriental).

14 I.A.1g(1) Latifoliados, dominados por dicotiledóneas.

15 I.A.1g(2) Dominados por palmeras, pero con árboles latifoliados en el sotobosque (p.e. pantanos de *Raphia taedigera* de Costa Rica).

16 I.A.1h *Bosque tropical sempervirente turboso.* (Con depósitos orgánicos en superficie.) Pobre en especies arbóreas, el dosel superior a menudo formado por un patrón de árboles altos en la periferia de la turba y más pequeños hacia el centro. Los árboles tienen diámetros pequeños y comúnmente provistos de neumatóforos o raíces zancos (fúlcreas).

I.A.2 BOSQUE TROPICAL Y SUBTROPICAL SEMPERVIRENTE ESTACIONAL. Compuesto principalmente por árboles sempervirentes con alguna protección en las yemas. Es posible observar una reducción del follaje en la estación seca, a menudo con defoliación parcial, este grupo de formaciones es transicional entre los grupos I.A.1 y I.A.3. Las subdivisiones a, b y c son bastante similares a las subdivisiones del bosque tropical ombrófilo (I.A.1).

17 I.A.2a *Bosque tropical (o subtropical) sempervirente estacional de baja altitud.*

I.A.2b *Bosque tropical (o subtropical) sempervirente estacional submontano.*

18 I.A.2b(1) Latifoliada.

19 I.A.2b(2) Aciculifoliadas.

20 I.A.2c *Bosque tropical (o subtropical) sempervirente estacional montano.* En contraste con I.A.1c no contiene helechos arborescentes; en lugar de éstos, arbustos sempervirentes son muy frecuentes.

21 I.A.2d *Bosque tropical (o subtropical) sempervirente seco subalpino.* Se parece fisionómicamente al bosque tropical con lluvias invernales (I.A.8a), generalmente se presenta más arriba del bosque nublado (I.A.1e). Mayormente con árboles sempervirentes esclerófilos, con menos de 20 metros, con poco o ningún sotobosque (si no ha sido abierto por la actividad humana), pobre en lianas y epífitas excepto líquenes.

I.A.3 BOSQUE TROPICAL Y SUBTROPICAL SEMIDECIDUO. La mayoría de los árboles del dosel superior son deciduos por sequía (drought deciduous); muchos de los árboles y arbustos de los estratos intermedios son sempervirentes y más o menos esclerófilos. Sin embargo, las plantas leñosas sempervirentes y deciduas no se encuentran netamente separadas por estratos, pueden presentarse mezcladas en un mismo estrato, o los arbustos pueden ser principalmente deciduos y los árboles sempervirentes. Casi todos los árboles tienen protección en sus yemas; las hojas no son largamente acuminadas. Los árboles muestran corteza rugosa, excepto las de algunos árboles de tronco abombado que pueden presentarse.

22 I.A.3a *Bosque tropical o subtropical semideciduo de baja altitud.* Los árboles más altos tienen a menudo el tronco abombado (p.e. *Ceiba*). Prácticamente no se presentan epífitas. El sotobosque está compuesto de regeneración de los árboles y de verdaderos arbustos leñosos. Pueden presentarse plantas suculentas (p.e. en forma de cactos cespitosos de tallo delgado); se presentan ocasionalmente lianas terofíticas y hemicriptofíticas. Puede presentarse un estrato poco denso de herbáceas compuesto principalmente de hemicroptófitas graminoides y de latifoliadas herbáceas.

23 I.A.3b *Bosque tropical (o subtropical) semideciduo montano o nublado.* Es similar a I.A.3a, el bosque semideciduo de baja altitud, pero el dosel superior es más bajo y está cubierto de epífitas xerofíticas (p.e. *Tillandsia usneoides*). Dentro del grupo I.A.3, de semideciduos no puede distinguirse una formación submontana.

I.A.4 BOSQUE SUBTROPICAL OMBRÓFILO. Se presenta sólo localmente en comunidades fragmentarias pequeñas, a causa de que el clima subtropical es típicamente un clima con estación seca. Donde se presenta el bosque ombrófilo subtropical, por ejemplo en Queensland (Australia) y en Formosa, cambia casi imperceptiblemente hacia el bosque tropical ombrófilo. El estrato inferior puede contener algunos arbustos. Sin embargo, el bosque subtropical ombrófilo no debe ser confundido con el bosque tropical ombrófilo montano el cual se presenta en un clima con temperatura media anual similar pero con diferencias térmicas entre verano e invierno menos pronunciadas. En consecuencia, los ritmos estacionales son más evidentes en todos los bosques subtropicales, aún en los ombrófilos.

El bosque subtropical ombrófilo está fisionómicamente más relacionado con el tropical que con el templado.

24 I.A.4a (Las subdivisiones se ajustan más o menos a las del bosque tropical ombrófilo, I.A.1a hasta h.)

25 I.A.4b

26 I.A.4c(1)
27 I.A.4c(2)
28 I.A.4c(3)
29 I.A.4c(4)

30 I.A.4d

31 I.A.4e(1)
32 I.A.4e(2)
33 I.A.4e(3)

34 I.A.4f(1)
35 I.A.4f(2)
36 I.A.4f(3)

37 I.A.4g(1)
38 I.A.4g(2)

39 I.A.4h

40 I.A.5 BOSQUE DE MANGLARES. Se presenta sólo en áreas con influencia de las mareas en zonas tropicales o subtropicales. Está compuesto casi únicamente de árboles y arbustos latifoliados esclerófilos sempervirentes, con zancos (raíces fúlcreas) o con neumatóforos, las epífitas son generalmente raras, excepto los líquenes en las ramas y algas adheridas en las partes bajas de los árboles. Es posible hacer subdivisiones; existen transiciones hacia el bosque tropical ombrófilo pantanoso [I.A.1g] (p.e. costas de Borneo, Nueva Guinea, etc.).

I.A.6 BOSQUE TEMPLADO Y SUBPOLAR SEMPERVIRENTE OMBRÓFILO. Se presenta sólo en climas extremadamente oceánicos casi libres de heladas, en el hemisferio sur principalmente en Chile. Se componen mayormente de árboles y arbustos verdaderamente sempervirentes hemiesclerófilos. Rico en talo-epífitas y en helechos herbáceos que enraizan en el suelo.

I.A.6a *Bosque templado latifoliado sempervirente ombrófilo.* Se presentan algunas lianas y epífitas vasculares; la altura generalmente es mayor de 10 metros.

41 I.A.6a(1) Latifoliados especialmente (p.e. bosques de *Nothofagus* de Nueva Zelandia).
42 I.A.6a(2) Con árboles aciculifoliados entremezclados.
43 I.A.6a(3) Principalmente acicufoliados o escuamifoliados (p.e. bosques de *Podocarpus* de Nueva Zelandia).

44	I.A.6b	*Bosque templado semper virente ombrófilo aluvial.* Rico en latifoliadas herbáceas (p.e. Nueva Zelandia occidental).
	I.A.6c	*Bosque templado semper virente ombrófilo pantanoso.*
45	I.A.6c(1)	Aciculifoliado o escuamifoliado. Bosque denso, alto (hasta 50 metros y más) escuamifoliado de baja altitud. Con frecuencia, raíces tabulares. Sotobosque denso a claro de graminoides (principalmente ciperáceas y plantas herbáceas latifoliadas (principalmente helechos). Rico en epífitas vasculares y brioides; algunas lianas (p.e. comunidades de *Bodocarpus dacrydioides* de Nueva Zelandia).
46	I.A.6c(2)	Latifoliados. Bosque alto (hasta 50 metros y más) latifoliados de baja altitud. Árboles espaciados de forma que se tocan las copas pero el dosel permite pasar mucha luz. Sinusia de arbustiva bastante clara. Faltan las epífitas en el dosel (p.e. bosques de *Eucalyptus ovata* de Victoria).
47	I.A.6d	*Bosque subpolar semper virente ombrófilo.* Se diferencia del bosque templado latifoliado (I.A.6a) porque aquí faltan las epífitas vasculares, y la altura del dosel superior es más reducida (en general menos de 10 metros). También el tamaño de las hojas es más reducido (p.e. bosques de hayas de Nueva Zelandia).
	I.A.7	BOSQUE TEMPLADO LATIFOLIADO SEMPERVIRENTE ESTACIONAL (CON LLUVIAS ADECUADAS EN VERANO). Compuesto principalmente de árboles y arbustos sempervirentes hemi-esclerófilos; sotobosque rico en caméfitas y hemicriptófitas herbáceas. Muy pocas o ninguna liana y epífitas vasculares. Varía hacia los bosques ombrófilos subtropicales (I.A.4) o templados (I.A.6) o hacia los bosques latifoliados sempervirente esclerófilos (I.A.8). Probablemente incluye tipos subpolares.
48	I.A.7a	Son posibles las subdivisiones en forma similar a las de los bosques tropicales y subtropicales estacionales, I.A.2a hasta d.)
49	I.A.7b(1)	
50	I.A.7b(2)	
51	I.A.7c	
52	I.A.7d	
	I.A.8	BOSQUE LATIFOLIADO SEMPERVIRENTE ESCLERÓFILO CON LLUVIAS DE INVIERNO. (A menudo se toma por mediterráneo, pero presente también en el sudoeste de Australia y en Chile, etc. El clima tiene una sequía de verano pronunciada.) Formado principalmente por árboles y arbustos sempervirentes esclerófilos, muchos de los cuales presentan corteza rugosa. Sotobosque herbáceo casi inexistente, no se presentan epífitas vasculares y solamente unas pocas epífitas criptogámicas, se presentan lianas leñosas sempervirentes.
53	I.A.8a	*Bosque latifoliado sempervirente esclerófilo con lluvias de invierno de baja altitud* (incluendo el submontano). Compuesto de eucaliptos gigantes (p.e. *Eucalyptus regnans* en Victoria y *E. diversicolor* en Australia occidental). Dominado por árboles mayores de 50 metros de altura.
54	I.A.8b	Ampliamente descrita en I.A.8 pero árboles menores de 50 metros de altura (p.e. bosques de encinas de California).
55	I.A.8c	*Bosque latifoliado sempervirente esclerófilo con lluvias de invierno, aluviales y pantanosos.* Este tipo tal vez existe, pero no es suficientemente conocido.
	I.A.9	BOSQUE SEMPERVIRENTE TROPICAL Y SUBTROPICAL DE CONÍFERAS. Formado principalmente por árboles acicufolados o escuamifoliados. Pueden existir árboles latifoliados entremezclados. Epífitas y lianas vasculares prácticamente inexistantes.

56 I.A.9a *Bosque sempervirente tropical y subtropical de coníferas, de baja altitud y submontanos* (p.e. bosque de pinos de Honduras y del Nicaragua).

57 I.A.9b *Bosque sempervirente tropical y subtropical montano y subalpino* (p.e. bosques de pinos de Filipinas y del Sur de México).

I.A.10 BOSQUE TEMPLADO Y SUBPOLAR SEMPERVIRENTE DE CONÍFERAS. Formado principalmente por árboles sempervirentes, aciculifoliados y escuamifoliados, pero pueden existir árboles latifoliados entremezclados. Epífitas y lianas vasculares prácticamente inexistentes.

58 I.A.10a *Bosque sempervirente de coníferas gigantes.* Dominado por árboles mayores de 50 metros de altura (p.e. bosques de *Sequoia* y de *Pseudotsuga* en el Pacífico Occidental de Norteamérica).

I.A.10b *Bosque sempervirente de coníferas con copas redondeadas.* Dominado por árboles de 5 hasta 50 metros de altura, con copas más o menos expandidas e irregularmente redondeadas (p.e. *Pinus* spp.).

59 I.A.10b(1) Con estratos intermedios sempervirentes esclerófilos (mediterráneos).

60 I.A.10b(2) Sin estratos intermedios sempervirentes esclerófilos.

61 I.A.10c *Bosque sempervirente de coníferas con copas cónicas.* Dominado por árboles de 5 hasta 50 metros de altura (sólo excepcionalmente más altos), con copas más o menos cónicas, como la mayoría de *Picea* y *Abies* (p.e. bosques de *Pseudotsuga Douglasii* de California).

62 I.A.10d *Bosque sempervirente de coníferas (no gigantes) con copas cilíndricas (boreal).* Similar a I.A.1c pero las copas tienen ramas muy cortas y por lo tanto son cilindro-cónicas, estrechas.

I.B *BOSQUE MAYORMENTE DECIDUO.* La mayoría de los árboles pierden su follaje simultáneamente y en conexión con la estación desfavorable.

I.B.1 BOSQUE DECIDUO POR LA SEQUÍA (TROPICAL Y SUBTROPICAL). La estación desfavorable está caracterizada principalmente por sequía, en la mayoría de los casos sequía de invierno. El follaje se pierde cada año. La mayoría de los árboles con corteza relativamente gruesa y fisurada.

63 I.B.1a *Bosque deciduo por sequía, de baja altitud (y submontano).* Prácticamente no hay plantas sempervirentes en ningún estrato, excepto algunas suculentas. Se presentan ocasionalmente lianas herbáceas y leñosas y también árboles de tronco abombado deciduos; la vegetación del piso principalmente de herbáceas (hemicriptófitas particularmente graminoides, geófitas y algunas terófitas), pero esparcidas (p.e. bosques deciduos de latifoliados del Noroeste de Costa Rica).

64 I.B.1b *Bosque deciduo por sequía, montano (y nublado).* Algunas especies sempervirentes en los estratos intermedios. Epífitas resistentes a la sequía presentes o abundantes, a menudo con forma barbada (por ejemplo: *Usnea* o *Tillandsia usneoides*). Esta formación no es frecuente pero es bien desarrollada, por ejemplo en el norte de Perú.

I.B.2 BOSQUE DECIDUO POR EL FRÍO CON ÁRBOLES (O ARBUSTOS) SEMPERVIRENTES ENTREMEZCLADOS. La estación desfavorable está caracterizada principalmente por heladas invernales. Los latifoliados deciduos son dominantes, pero especies sempervirentes se presentan formando parte del dosel principal o de los estratos intermedios. Trepadoras y epífitas vasculares, escasas o ausentes.

65 I.B.2a *Bosque deciduo por el frío, con árboles y trepadoras latifoliadas sempervirentes* (p.e. *Ilex aquifolium* y *Hedera helix* en Europa Occidental). Rico en epífitas criptogámicas

		incluyendo musgos. Se pueden presentar epífitas vasculares en la base de los troncos. Lianas a veces comunes en los llanos de inundación.
66	I.B.2b	*Bosque deciduo por el frío, con árboles latifoliados sempervirentes* (p.e. bosques de arces y de *Tsuga* de Nueva York).
	I.B.3	BOSQUE DECIDUO POR EL FRÍO SIN ÁRBOLES SEMPERVIRENTES. Dominan absolutamente los árboles deciduos. Pueden presentarse caméfitas sempervirentes y algunas manofanerófitas sempervirentes. Las trepadoras de presencia insignificante, (pero pueden ser comunes en los llanos de inundación); epífitas vasculares ausentes (excepto ocasionalmente en el pie de los árboles); las taloepífitas, especialmente los líquenes, están siempre presentes.
67	I.B.3a	*Bosque templado deciduo por el frío, de baja altitud submontano.* Los árboles alcanzan hasta 50 metros de altura (p.e. bosque mesofítico entremezclado de los Estados Unidos de América.) Las epífitas son principalmente algas y líquenes costrosos.
	I.B.3b	*Bosque deciduo por el frío, montano o boreal* (incluyendo bosques de baja altitud o submontano influenciados por situación topográfica con alta humedad atmosférica). Las epífitas son líquenes foliosos y fruticosas; briófitas. Los árboles alcanzan hasta 50 metros de altura, pero en bosques montañosos o boreales normalmente no son mayores de 30 metros.
68	I.B.3b(1)	Principalmente latifoliados deciduos.
69	I.B.3b(2)	Principalmente aciculifoliados deciduos (p.e. *Larix* de Siberia).
70	I.B.3b(3)	Latifoliados y aciculifoliados mezclados.
	I.B.3c	*Bosque deciduo por el frío, subalpino o polar.* En contraste con las formaciones I.B.3a y b, estos bosques tienen el dosel superior a una altura significativamente reducida (no sobrepasan los 20 metros). Los troncos frecuentemente son retorcidos. Las epífitas son similares a las de la formación b pero en general más abundantes. A menudo cambian hacia bosques claros (véase clase II).
71	I.B.3c(1)	Con sotobosque principalmente de hemicriptófitas.
72	I.B.3c(2)	Con sotobosque principalmente de caméfitas. Su identidad puede camuflarse con bosques entremezclados de coníferas.
	I.B.3d	*Bosque deciduo por el frío, aluvial.* Inundado por cursos de agua, por lo tanto las áreas donde se localizan son más húmedas y más ricas en nutrientes que aquéllas de la formación I.B.3a, bosque deciduo por el frío, de baja altitud. Árboles y arbustos de crecimiento rápido y herbáceas vigorosas constituyen el sotobosque.
73	I.B.3d(1)	Ocasionalmente inundados, fisionómicamente similar a la formación I.B.3a, con sotobosque de árboles altos y abundantes plantas arbustivas macrófilas.
74	I.B.3d(2)	Periódicamente inundados. Los árboles no son tan altos ni tan densos como en I.B.3a, pero el sotobosque de herbáceas es abundante y alto (en Eurasia dominan frecuentemente especies de *Salix* y *Alnus*).
	I.B.3e	*Bosque deciduo por el frío, pantanoso o turboso.* Inundado hasta finales de primavera o principios de invierno, la superficie del suelo es orgánica. Relativamente pobre en especies arbóreas, la cobertura del piso con formas de vida variadas mayormente continuo.
75	I.B.3e(1)	(Subdivisiones semejantes a las de I.B.3b, bosque deciduo por el frío, boreal.)
76	I.B.3e(2)	
77	I.B.3e(3)	
	I.C	*BOSQUES EXTREMADAMENTE XEROMÓRFICOS.* Los rodales son densos de fanerófitas xeromórficas, tales como las de árboles de tronco abombado,

árboles estipitados con hojas y tallos suculentos. El sotobosque de arbustos con adaptaciones xeromórficas similares, caméfitas y hemicriptófitas herbáceas, geófitas y terófitas. A menudo su transición es hacia bosques claros (véase clase II).

78 I.C.1 — BOSQUE EXTREMADAMENTE XEROMÓRFICO DOMINADO POR ESCLERÓFILAS. Combinación de formas de vida como en el anterior, excepto por la predominancia de árboles esclerófilos, muchos de los cuales tienen la base del tallo bulbosa, profundamente enterrados en el suelo (xilópodos).

I.C.2 — BOSQUE ESPINOSO. Predominan las especies con apéndices espinosos.

79 I.C.2a — *Bosque espinoso deciduo-sempervirente (mixto).*

80 I.C.2b — *Bosque espinoso enteramente deciduo.*

81 I.C.3 — BOSQUE MAYORMENTE SUCULENTO. Son muy frecuentes las plantas suculentas con forma arbórea (escaposas) y formas arbustivas (cespitosas) pero también se presentan las otras xerofanerófitas.

II BOSQUE CLARO (comunidad de árboles abierta)

Formado por árboles de por lo menos 5 metros de altura, la mayoría de las copas no se tocan entre ellas, pero cubren por lo menos el 40% de la superficie. Puede existir una sinusia herbácea. Véase el grupo de formaciones V.A.1 si la abundancia de árboles es menor del 40% y hay una sinusia herbácea. El límite del 40% es conveniente porque puede calcularse fácilmente durante el trabajo sobre el terreno: cuando la abundancia de árboles es del 40%, la distancia entre dos copas es igual al radio medio de una copa.

II.A *BOSQUE CLARO MAYORMENTE SEMPERVIRENTE.* Sempervirente definido en I.A.

82 II.A.1 — BOSQUE CLARO LATIFOLIADO SEMPERVIRENTE. Mayormente los árboles y arbustos esclerófilos, no hay epífitas.

II.A.2 — BOSQUE CLARO ACICULIFOLIADO SEMPERVIRENTE. Principalmente aciculifoliado o escuamifoliado. Las copas de muchos árboles se extienden hasta la base de los tallos, o son por lo menos muy ramificadas.

II.A.2a — *Bosque claro aciculifoliado sempervirente (de coníferas) con copas redondeadas* (p.e. *Pinus*).
83 II.A.2a(1) — Con estratos intermedios sempervirentes esclerófilos (Mediterráneo).
84 II.A.2a(2) — Sin estratos intermedios sempervirentes esclerófilos.

85 II.A.2b — *Bosque claro aciculifoliado sempervirente prevaleciendo las copas cilíndricas.* (En su mayoría subalpino.)

86 II.A.2c — *Bosque claro de coníferas sempervirentes con copas cilindro-cónicas muy angostas* (p.e. *Picea* en la región boreal).

II.B *BOSQUE CLARO MAYORMENTE DECIDUO* (véase I.B).

II.B.1 — BOSQUE CLARO DECIDUO POR LA SEQUÍA

87 II.B.1a — (Subdivisiones más o menos semejantes a las de los bosques densos, I.B.1.)

88 II.B.1b

	II.B.2	BOSQUE CLARO DECIDUO POR EL FRÍO, CON ÁRBOLES SEMPERVIRENTES
89	II.B.2a	(Subdivisiones semejantes a las de I.B.2.)
90	II.B.2b	
	II.B.3	BOSQUE CLARO DECIDUO POR EL FRÍO, SIN ÁRBOLES SEMPERVIRENTES (véase I.B.3). Muy frecuentes en la región subártica, en otras partes solamente en pantanos o ciénagas.
91	II.B.3a	*Bosque claro latifoliado deciduo.*
92	II.B.3b	*Bosque claro aciculifoliado deciduo.*
93	II.B.3c	*Bosque claro mixto deciduo (latifoliado y aciculifoliado).*
	II.C	*BOSQUE CLARO EXTREMADAMENTE XEROMÓRFICO.* Similar al I.C la única diferencia es un menor entrelazamiento de los árboles individuales.
94	II.C.1	(Subdivisiones como en I.C.)
95	II.C.2a	
96	II.C.2b	
97	II.C.3	
	III	MATORRAL (matorral denso o claro)
		Mayormente constituido por fanerófitas leñosas cespitosas de 0,5 a 5 metros de altura [1]. Cada una de las siguientes subdivisiones puede ser aplicada a: Matorral claro. Los arbustos no se tocan entre sí; frecuentemente con un estrato graminoide. Matorral denso. Arbustos individuales entrelazados.
	III.A	*MATORRAL MAYORMENTE SEMPERVIRENTE.* (Sempervirente en el sentido de I.A.)
	III.A.1	MATORRAL CLARO LATIFOLIADO SEMPERVIRENTE (O MATORRAL DENSO).
98	III.A.1a	*Matorral denso bajo de bambú* (o, menos frecuentemente, matorral claro). Macro- o micro-fanerófitas graminoides lignificadas reptantes.
99	III.A.1b	*Matorral claro con árboles estipitados sempervirentes (o matorral denso).* Compuesto de árboles pequeños y arbustos pequeños (p.e. matorral claro de palmas enanas del Mediterráneo o matorral denso de helechos arborescentes de Hawaii).
100	III.A.1c	*Matorral denso latifoliado sempervirente esclerófilo (o matorral claro).* Macro- o micro-fanerófitas apoyadas entre sí, o cespitosas reptantes con hojas relativamente grandes y suaves (p.e. matorrales densos subalpinos de *Rhododendron*, o matorrales densos de *Hibiscus tileaeceus* de Hawaii).
101	III.A.1d	*Matorral claro latifoliado sempervirente esclerófilo (o matorral denso).* Dominado por arbustos esclerófilos y árboles inmaduros (p.e. Chaparral o macchia). A menudo puede camuflarse con vegetación abierta tipo parque, vegetación graminoide herbácea, o brezales.

1. No debe ser confundido con bosque secundario en desarrollo (véase nota, p. 71). Algunas veces el matorral puede alcanzar más de 5 metros de altura.

102 III.A.1e *Matorral denso latifoliado sufruticoso (o matorral claro).* Comunidad de nanofanerófitas semilignificadas que en años secos pueden perder las hojas en parte de su sistema de brotes (p.e. brezal de *Cistus*) [1].

III.A.2 MATORRAL CLARO SEMPERVIRENTE ACICULIFOLIADO MICRÓFILO (O MATORRAL DENSO).

103 III.A.2a *Matorral denso sempervirente aciculifoliado (o matorral claro).* Compuesto en su mayoría de fanerófitas aciculifoliadas apoyadas entre sí o reptantes (p.e. *Pinus mughus* "Krummholz").

104 III.A.2b *Matorral claro sempervirente micrófilo (o matorral denso).* A menudo arbustos ericoideos (en su mayoría se presentan en pisos altitudinales subalpinos tropicales).

III.B. *MATORRAL MAYORMENTE DECIDUO.* (Deciduos en el sentido de I.B.)

105 III.B.1 MATORRAL DECIDUO POR LA SEQUÍA, CON PLANTAS LEÑOSAS SEMPERVIRENTES ENTREMEZCLADAS.

106 III.B.2 MATORRAL DECIDUO POR LA SEQUÍA, SIN PLANTAS LEÑOSAS SEMPERVIRENTES ENTREMEZCLADAS.

III.B.3 MATORRAL DECIDUO POR EL FRÍO.

107 III.B.3a *Matorral denso templado deciduo (o matorral claro).* Es un matorral más o menos denso sin o con muy poco sotobosque de herbáceas pequeñas. Pobre en criptógamas.

III.B.3b *Matorral denso subalpino o subpolar deciduo (o matorral claro).* Nanofanerófitas cespitosas apoyadas entre sí o erectas con gran capacidad de regeneración vegetativa. Por regla general está cubierto de nieve por lo menos la mitad del año.

108 III.B.3b(1) Principalmente con sotobosque de hemicriptófitas, más frecuentemente latifoliadas herbáceas (p.e. matorral denso subalpino de *Alnus viridis*).

109 III.B.3b(2) Principalmente con sotobosque de caméfitas, más frecuentemente arbustos enanos y líquenes fruticosos (p.e. matorral claro de *Betula tortuosa* en el límite latitudinal de árboles).

III.B.3c *Matorral claro deciduo aluvial (o matorral denso).* Arbustos de crecimiento rápido que se presentan como pioneros en bancos de cursos de agua o en islas que a menudo son intensamente inundadas; por lo tanto, la mayoría con sotobosque muy esparcido.

110 III.B.3c(1) Con hojas lanceoladas (p.e. *Salix*, en su mayoría se presentan en piso submontano o en bajura).

111 III.B.3c(2) Micrófilos.

112 III.B.3d *Matorral claro deciduo turboso (o matorral denso).* Nanofanerófitas cespitosas erectas con *Sphagnum* y (u) otros musgos turbosos.

III.C *MATORRAL CLARO EXTREMADAMENTE XEROMÓRFICO (SUBDESIERTO).* Comunidades muy abiertas de arbustos con varias adaptaciones xerofíticas, tales como ser extremadamente escleromórficas o con hojas muy reducidas, o con ramas verdes y sin hojas, o tallos suculentos, etc., algunos con espinas.

III.C.1 MATORRAL CLARO SUBDESÉRTICO MAYORMENTE SEMPERVIRENTE. En años extremadamente secos pueden perder algunas hojas y porciones de brotes.

1. Ocasionalmente con menos de 50 cm de altura, por lo tanto es una transición hacia el IV.A.1a.

III.C.1a *Matorral claro subdesértico semperviriente.*

113 III.C.1a(1) Latifoliados dominados por nanofanerófita, esclerófitas, incluyendo algunos arbustos con filoclados (p.e. matorral de "mulga" en Australia).

114 III.C.1a(2) Micrófilos o áfilos, pero con tallos verdes (p.e. *Retama retam*).

115 III.C.1a(3) Suculentos, dominados por plantas suculentas con hojas y tallos diversamente ramificados.

III.C.1b *Matorral claro subdesértico semi-deciduo.* Con arbustos ya sean facultativamente deciduos o una combinación de semperviriente y deciduos.

116 III.C.1b(1) Facultativamente deciduos (p.e. matorrales salados [saltbrush] de *Atriplex* y *Kochia* en Australia).

117 III.C.1b(2) Mixtos deciduos y sempervirentes, es transicional a III.C.2.

III.C.2 MATORRALES CLAROS SUBDESÉRTICOS DECIDUOS. Arbustos mayormente deciduos a menudo con unos pocos sempervirentes.

118 III.C.2a *Matorral claro subdesértico deciduo sin suculentas.*

119 III.C.2b *Matorral claro subdesértico deciduo con suculentas.*

IV MATORRAL ENANO Y COMUNIDADES RELACIONADAS

Rara vez sobrepasan los 50 cm de altura (algunas veces son llamados brezales o formaciones brezaloides).

Según la densidad de cobertura de los arbustos enanos se pueden distinguir:

Matorral denso enano: ramas entrelazadas;

Matorral claro enano: arbustos enanos más o menos aislados o en grupos;

Formaciones criptogámicas con arbustos enanos: la superficie del suelo está densamente cubierta de musgos o líquenes (talo-caméfitas); se presentan arbustos enanos individualmente o en grupos pequeños; en el caso de ciénagas pueden incluirse comunidades graminoides dominantes.

IV.A *MATORRAL ENANO MAYORMENTE SEMPERVIRENTE.* En su mayoría arbustos enanos sempervirentes.

IV.A.1 MATORRAL DENSO ENANO SEMPERVIRENTE. Cobertura de arbustos enanos muy cerrada, dominando el paisaje (es el brezal enano en el sentido estricto).

120 IV.A.1a *Matorral denso enano semperviriente cespitoso.* La mayoría de las ramas permanecen en posición erecta, y a menudo cubiertas por líquenes foliosos. Musgos pulvinados en el piso, líquenes fruticosos o formas de vida herbáceas pueden tener importancia (p.e. brezal de *Calluna*).

121 IV.A.1b *Matorral denso enano semperviriente reptante o enmarañado.* La mayoría de las ramas se enmarañan sobre el piso. Diversamente combinados con talo-caméfitas entre las cuales pueden introducirse las ramas (p.e. brezal de *Loiseleuria*).

IV.A.2 MATORRAL CLARO ENANO SEMPERVIRENTE. Cobertura abierta o poco densa de los arbustos enanos.

122 IV.A.2a *Matorral claro de arbustos almohadillados sempervirentes.* Grupos de arbustos enanos más o menos aislados formados por almohadillas densas, frecuentemente equipados de espinas (p.e. brezal de "porcupine" de *Astragalus* y *Acantholimon* de las montañas del Mediterráneo oriental).

IV.A.3 MATORRAL DE ARBUSTOS ENANOS SEMPERVIRENTES Y HERBÁCEAS. Comunidades más o menos abiertas de caméfitas sempervirentes, sufruticosas o herbáceas, varias hemicriptófitas, geófitas, etc.

123 IV.A.3a *Matorral de arbustos enanos sempervirentes y formación mixta de hierbas* (p.e. brezal *Nardus-Calluna*).

124 IV.A.3b *Matorral de arbustos enanos parcialmente sempervirente y formación mixta de hierbas.* Muchos individuos pierden parte de sus brotes durante la estación seca (p.e. *Phrygana en* Grecia).

IV.B *MATORRAL ENANO MAYORMENTE DECIDUO.* Similares a IV.A pero en su mayoría formados por especies deciduas.

125 IV.B.1 MATORRAL DENSO ENANO O MATORRAL CLARO ENANO FACULTATIVAMENTE DECIDUOS POR LA SEQUÍA. El follaje se pierde sólo en años extremos.

IV.B.2 MATORRAL DENSO ENANO O MATORRAL CLARO ENANO (OBLIGATORIAMENTE) DECIDUOS POR LA SEQUÍA. Comunidades de arbustos enanos muy densos que pierden todas o parte de sus hojas en la estación seca.

126 IV.B.2a *Matorral denso enano cespitoso deciduo por la sequía.* Corresponde a IV.A.1a.

127 IV.B.2b *Matorral denso enano deciduo por la sequía, reptante o enmarañado.* Corresponde a IV.A.1b.

128 IV.B.2c *Matorral claro enano deciduo por la sequía con arbustos almohadillados.* Corresponde a IV.A.2a.

129 IV.B.2d *Matorral claro enano deciduo por la sequía en mosaico (o mixto).* Arbustos enanos deciduos y sempervirentes, hemicriptófitas cespitosas, caméfitas suculentas y otras formas de vida entremezcladas en varios patrones.

IV.B.3 MATORRAL DENSO ENANO O MATORRAL CLARO ENANO DECIDUOS POR EL FRÍO. Fisionómicamente similares a IV.B.2, pero pierden las hojas al principio de la estación fría. Generalmente más ricos en caméfitas criptogámicas.

130 IV.B.3a (Subdivisiones similares a IV.B.2; son posibles transiciones hacia IV.D y IV.E.)

131 IV.B.3b

132 IV.B.3c

133 IV.B.3d

IV.C *MATORRAL CLARO ENANO EXTREMADAMENTE XEROMÓRFICO.* Formación más o menos abierta compuesta de arbustos enanos, plantas suculentas, geófitas y otras formas de vida adaptadas para sobrevivir o para evitar los efectos de una estación seca larga. En su mayoría subdesértico.

134 IV.C.1a(1) (Subdivisiones semejantes a III.C.)
135 IV.C.1a(2)
136 IV.C.1a(3)

137 IV.C.1b(1)
138 IV.C.1b(2)

139 IV.C.2a
140 IV.C.2b

IV.D *TUNDRA DE ARBUSTOS ENANOS, LÍQUENES Y MUSGOS.* Formaciones bajas de crecimiento lento compuestas principalmente de arbustos enanos, de graminoides y criptógamas, localizadas fuera del límite latitudinal de los árboles (regiones subpolares). A menudo se presentan patrones de distribución causados por movimientos del suelo por congelación (cryoturbation). Excepto en regiones boreales, las formaciones de arbustos enanos que crecen sobre el límite altitudinal de los árboles, no deberán llamarse tundras, ya que por regla general ellas son ricas en arbustos enanos y graminoides, y crecen más altas debido a la mayor radiación de las bajas latitudes.

IV.D.1 TUNDRA MAYORMENTE DE BRIÓFITAS. Dominada por un tapizado o pequeñas almohadillas de musgos camefíticos. Por regla general hay grupos de arbustos enanos y distribuidos regularmente y no muy densos. El aspecto general es más o menos verde obscuro, verde oliva o pardusco.

141 IV.D.1a *Tundra de musgos y arbustos enanos cespitosos.*

142 IV.D.1b *Tundra de musgos y arbustos enanos reptantes o enmarañados.*

143 IV.D.2 TUNDRA MAYORMENTE DE LÍQUENES. Domina un tapiz de líquenes fruticosos, dando a la formación un aspecto más o menos gris oscuro. Los arbustos enanos mayormente sempervirentes, reptantes o pulcrinados. Tundra de líquenes y arbustos enanos.

IV.E *FORMACIÓN DE PANTANO TURBOSO DE MUSGOS CON ARBUSTOS ENANOS.* Existen acumulaciones de turba oligomórfica constituida principalmente de *Sphagnum* u otros musgos, los cuales por lo general cubren también el piso. Los arbustos enanos están concentrados en las partes relativamente más secas o bastante dispersos. Se asemejan desde cierto punto de vista a las formaciones de arbustos enanos de suelos minerales. Localmente pueden predominar geófitas con rizomas, hemicriptófitas graminoides y otras formas de vida herbáceas, algunos árboles y arbustos de crecimiento lento pueden crecer como individuos aislados, o en grupos, o en bosques claros, los cuales son marginales al pantano turboso o pueden ser reemplazados por formaciones abiertas en una sucesión cíclica. Las siguientes subdivisiones corresponden a la clasificación de pantanos turbosos adoptada en Europa.

IV.E.1 PANTANO TURBOSO ELEVADO. Por el crecimiento de especies de *Sphagnum* se elevan sobre la napa freática general y tienen una napa freática superficial propia. Por lo tanto no reciben agua "mineral" (es decir, agua que haya estado en contacto con el suelo inorgánico), sino solamente agua de lluvia (pantanos musgosos verdaderamente ombrofíticos).

144 IV.E.1a *Pantano turboso elevado típico (suboceánico, de baja altitud y submontano).* Dominan los musgos en toda la extensión excepto en los montículos secos elevados localmente, los cuales son dominados por arbustos enanos. Los árboles son raros y si están presentes se concentran en las pendientes marginales de la acumulación turbosa convexa. La mayoría está rodeada de un pantano muy húmedo y poco oligotrófico.

145 IV.E.1b *Pantano turboso elevado montano (o subalpino).* Crece más lento que el pantano turboso elevado típico (o fue formado en un periodo antiguo con un clima más caliente y actualmente ha "muerto" o ha sido destruido por erosión). Frecuentemente está cubierto de juncos o arbustos enanos sempervirentes. Localmente pueden dominar las micro- o nano-fanerófitas (p.e. *Pinus mughus*).

146 IV.E.1c *Pantano turboso de bosque claro subcontinental.* Temporalmente cubierto de bosque claro de baja productividad, el cual en una secuencia de años más húmedos puede ser reemplazado por formaciones de *Sphagnum* semejantes a E.1a.

IV.E.2 PANTANOS TURBOSOS NO ELEVADOS. No están, o lo están poco elevados sobre la napa freática de agua mineral que rodea el paisaje. Por lo tanto es más húmedo y menos oligomórfico que IV.E.1. Más pobre en musgos que el pantano elevado típico (IV.E.1a), con el cual pueden presentarse varias formas de transición.

147 IV.E.21 *Pantano turboso tapizado de baja altitud, oceánico, submontano, o montano.* La microsuperficie del pantano turboso es menos ondulada y menos rica en musgos de crecimiento activo, que el IV.E.1a. Arbustos enanos semperviréntes están esparcidos y también hemicritófitas cespitosas (juncos o graminoides) y algunas geófitas rizomatosas

148 IV.E.2b *Pantano turboso acordoneado (pantano turboso "Aapa" de Finlandia).* Medio oligótrofo plano con montículos en forma de cordones en las bajuras boreales. El nombre finlandés indica un pantano turboso abierto sin o con muy pocos árboles de muy poco vigor, los cuales crecen en montículos alargados bajos y delgados, llamados "cordones"; estos cordones turbosos se forman por la presión del hielo que cubre más o menos los pantanos turbosos inundados desde el principio hasta el fin de la primavera. Solamente estos cordones están cubiertos de arbustos enanos y son ricos en *Sphagnum*. La parte principal del pantano es similar a un pantano de juncos húmedo.

Vegetación herbácea

La clasificación de la vegetación herbácea requiere especial atención principalmente a causa de: *a*) los cambios estacionales continuos en la fisionomía de las formaciones; *b*) los problemas que se plantean a menudo para distinguir las formaciones herbáceas tropicales de las no tropicales; *c*) la explotación de las formaciones herbáceas que puede afectar profundamente a la estructura de la vegetación y que puede cambiar con frecuencia; *d*) las dificultades para distinguir entre formaciones herbáceas naturales y artificiales.

Existen dos formas principales de crecimiento herbáceo: graminoides y forbias [1]. Las graminoides incluyen todas las plantas herbáceas semejantes a las gramíneas, tales como las ciperáceas (*Carex*), las juncáceas (*Juncus*), las tafáceas (*Typha*), etc. Entre las forbias figuran el trébol (*Trifolium*), el girasol (*Helianthus*), los helechos, las *Asclepias*, etc. En general, se incluyen en las forbias todas las plantas herbáceas no graminoides.

En la clasificación de la Unesco se emplea con frecuencia el término de formación graminoide para designar un tipo de vegetación dominado por formas de crecimiento análogo de las gramíneas. Como los graminoides incluyen numerosos taxones distintos de las gramíneas, el término formación graminoide debe interpretarse aquí como un tipo de vegetación fisionómica sin significación florística.

Lo mismo que en las formaciones leñosas, la altura constituye un rasgo importante para caracterizar una comunidad vegetal dominada por formas de crecimiento herbáceo. Las grandes fluctuaciones estacionales en la altura de las plantas herbáceas requieren que ese carácter se mida (o se estime) en el momento de la floración, es decir, cuando se han desarrollado las inflorescencias. Éstas pueden no desarrollarse donde prevalece el pastoreo constante o intensivo y en ese caso hay que estimar la altura.

La cobertura es otro factor importante para caracterizar la vegetación herbácea. Sin embargo, a la escala de 1/1 000 000 o menor, se supone que todas las formaciones herbáceas tienen una cobertura más o menos continua, y no se hace mención de ella en la leyenda del mapa. La excepción es una densidad muy baja. En ese caso, la formación se denomina abierta.

Las formas de crecimiento de las graminoides tienen importancia y es corriente establecer una distinción entre formas cespitosas y amacolladas (montículos) o combinaciones de ambas. En la clasificación de la Unesco, se considera que todas las comunidades graminoides son más o menos cespitosas y, por consiguiente, no hace falta mencionarlo. Sin embargo, el predominio de formas amacolladas afecta profundamente a la fisionomía de la vegetación. Por consiguiente, la forma ama-

1. Neologismo que hemos derivado del griego *forbe* — pasto — por analogía con el inglés *forbs*. En español se emplea ya la palabra "euforbias" de la misma etimología para un grupo particular de plantas.

collada debe indicarse en la descripción de las comunidades vegetales de que se trate. Esas descripciones son paralelas a las de las formas de las copas de los árboles sempervirentes aciculifoliados, como por ejemplo en I.A.9c y d.

Los tipos de vegetación herbácea comprenden a menudo una sinusia de plantas leñosas que comunica un carácter especial. Por eso, con frecuencia se utilizan esas sinusias para caracterizar las comunidades herbáceas. Entre las características más importantes de la sinusia leñosa figuran la altura y la densidad, así como si son sempervirentes o deciduas, aciculifoliadas, latifoliadas o afoliadas (esencialmente sin hojas), etc. Esas características se presentan en gran variedad de combinaciones y determinan gráficamente la fisionomía de la vegetación. Son admisibles otros rasgos si caracterizan a una formación lo bastante extensa para figurar en el mapa a escala 1/1 000 000 o menos, por ejemplo, la naturaleza "esclerófila" de la sinusia leñosa de muchas combinaciones de *Eucalyptus* y gramíneas en Australia.

Existen numerosos términos geobotánicos utilizados con frecuencia, tales como sabana, estepa, pradera, etc., que se han evitado en las definiciones de la clasificación de la Unesco porque se prestan a interpretaciones demasiado contradictorias. No obstante y ocasionalmente se han añadido entre paréntesis cuando contribuyen a que el lector identifique la categoría. Es siempre preferible utilizar la terminología de la Unesco en los mapas de vegetación. Pueden añadirse términos locales que tienen un sentido para los habitantes de las regiones respectivas (por ejemplo, campo cerrado) pero no sustituir con ellos a los términos de la Unesco en las leyendas de los mapas. De esa forma, los mapas de vegetación serán comprensibles tanto para los usuarios locales como para el público de todo el mundo. Esto es especialmente importante en los estudios comparados.

V VEGETACIÓN HERBÁCEA

V.A — *VEGETACIÓN GRAMINOIDE ALTA.* En las praderas graminoides altas, las formas del crecimiento graminoide dominante alcanzan más de 2 metros de altura cuando están plenamente desarrolladas las inflorescencias. Pueden existir forbias pero su cobertura es inferior al 50%.

V.A.1 — VEGETACIÓN GRAMINOIDE ALTA CON UNA SINUSIA ARBÓREA QUE CUBRE DEL 10 AL 40%, con o sin arbustos. Constituyen una especie de bosque muy abierto con una cobertura del suelo más o menos continuo (superior al 50%) de graminoides altas. Para las categorías con una sinusia de árboles que cubre más del 40% véase la clase de formaciones II.

149 V.A.1a — *Sinusia leñosa latifoliada sempervirente.*

150 V.A.1b — *Sinusia leñosa latifoliada semi-sempervirente,* es decir, compuesta por lo menos de un 25% de árboles latifoliados sempervirentes y otro 25% de árboles latifoliadas deciduos.

V.A.1c — *Sinusia leñosa latifoliada decidua.*

151 V.A.1c(1) — Similar a la anterior pero inundada estacionalmente, por ejemplo, en el noreste de Bolivia.

152 V.A.2 — VEGETACIÓN GRAMINOIDE ALTA CON UNA SINUSIA LEÑOSA QUE CUBRE MAS DEL 10%, con o sin arbustos.

V.A.2a — (Subdivisiones como en V.A.1.)

153 V.A.2b

154 V.A.2c

155 V.A.2d — *Formaciones graminoide alta tropical o subtropical con árboles o arbustos, que crecen en macollas sobre las termiteras* (sabana termitosa).

V.A.3 VEGETACIÓN GRAMINOIDE ALTA CON UNA SINUSIA DE ARBUSTOS (sabana arbustiva).

156 V.A.3a (Subdivisiones como en V.A.2.)

157 V.A.3b

158 V.A.3c

159 V.A.3d

V.A.4 VEGETACIÓN GRAMINOIDE ALTA CON UNA SINUSIA LEÑOSA COMPUESTA PRINCIPALMENTE DE PLANTAS EMPENACHADAS (por lo general palmeras).

160 V.A.4a *Formación graminoide tropical con palmeras* (p.e. las sabanas de palmeras de *Arocomia totai* y *Attalea princeps* al norte de Santa Cruz de la Sierra, Bolivia).

161 V.A.4a(1) Similares a las anteriores, estacionalmente inundadas (p.e. las sabanas de *Mauritia vinifera*, Llanos de Mojos, Bolivia).

V.A.5 VEGETACIÓN GRAMINOIDE ALTA, PRÁCTICAMENTE SIN SINUSIA LEÑOSA.

162 V.A.5a *Formación graminoide tropicale* como las descritas en V.A.5 en diferentes regiones de baja latitud de África.

163 V.A.5a(1) Similares a las anteriores, estacionalmente inundadas (p.e. los Campos de Várzea del valle inferior del Amazonas).

164 V.A.5a(2) Similares a las anteriores, húmedas o inundadas la mayor parte del año (p.e. las marismas de papirus, *Cyperus papyrus* del valle superior del Nilo).

V.B *VEGETACIÓN GRAMINOIDE DE ALTURA INTERMEDIA*, es decir, con las formas de crecimiento graminoide dominantes de 50 cm a 2 metros de altura cuando sus inflorescencias están plenamente desarrolladas. Puede haber forbias pero con una cobertura inferior al 50%. (Divisiones y sybdivisiones como V.A.1 a V.A.3d.)

165 V.B.1a

166 V.B.1b

167 V.B.1c

168 V.B.2a

169 V.B.2b

170 V.B.2c

171 V.B.2d

172 V.B.3a

173 V.B.3b

174 V.B.3c

175 V.B.3d

176 V.B.3e *Sinusia leñosa compuesta principalmente de arbustos espinosos deciduos* (p.e. la sabana tropical de arbustos espinosos de la región del Sahel en África con *Acacia tortilis, A. senegal y otras*).

V.B.4 VEGETACIÓN GRAMINOIDE DE ALTURA INTERMEDIA CON UNA SINUSIA ABIERTA DE PLANTAS EMPENACHADAS, por lo general palmeras.

177	V.B.4a	*Formación graminoide subtropical de altura intermedia con grupos claros de palmeras* (p.e. en Corrientes, Argentina).
178	V.B.4a(1)	Similares a las anteriores, estacionalmente inundadas (p.e. los bosquecillos de palmera *Mauritia* en los llanos colombianos y venezolanos).
	V.B.5	VEGETACIÓN GRAMINOIDE DE ALTURA INTERMEDIA PRÁCTICAMENTE SIN SINUSIA LEÑOSA.
179	V.B.5a	*Formación graminoide de altura intermedia compuesta principalmente de gramíneas cespitosas* (p.e. la pradera de gramíneas altas de Kansas Oriental).
180	V.B.5a(1)	Húmeda o inundada la mayor parte del año (p.e. las formaciones pantanosas de *Typha*).
181	V.B.5a(2)	Sobre suelo arenoso o dunas (p.e. las comunidades de *Andropogon hallii* en las dunas de arena de Nebraska).
182	V.B.5b	*Formación herbácea de altura intermedia compuesta principalmente de gramíneas amacolladas* (p.e. las praderas de *Festuca novaezelandiae* en Nueva Zelandia).
	V.C	*VEGETACIÓN GRAMINOIDE BAJA*, es decir, las formas de crecimiento graminoide dominantes son menores de 50 cm de altura cuando sus florescencias están desarrolladas. Puede haber forbias pero cubren menos del 50%. (Divisiones y subdivisiones como B.V.1 a V.B.4.)
183	V.C.1a	
184	V.C.1b	
185	V.C.1c	
186	V.C.2a	
187	V.C.2b	
188	V.C.2c	
189	V.C.2d	
190	V.C.3a	
191	V.C.3b	
192	V.C.3c	
193	V.C.3d	
194	V.C.3e	
195	V.C.4a	
196	V.C.4a(1)	
197	V.C.5a	*Comunidades alpinas tropicales abiertas* de gramíneas amacolladas de claras a densas con una sinusia leñosa de plantas empenachadas (*Espeletia*, *Lobelia*, *Senecio*), matas micrófilas a leptófilas y plantas almohadilladas, a menudo con hojas algodonosas. Por encima del límite de los árboles en las bajas latitudes: páramos y tipos de vegetación afines en las regiones alpinas sin nieve de Kenia, Colombia, Venezuela, etc.
198	V.C.5b	Similares a V.C.5a pero muy claras y sin plantas empenachadas. Frecuentemente nevadas nocturnas (la nieve desaparece hacia las nueve de la mañana); superpáramo (es decir, por encima del páramo) de J. Cuatrecasas [1].

1. "Páramo vegetation and its life forms", *Colloquium geographicum*, vol. 9, Bonn, F. Dümmlers Verlag, 1968, 223 p.

199 V.C.5c *Vegetación alpina tropical o subtropical de gramíneas amacolladas* con grupos claros de plantas sempervirentes, con o sin arbustos o matas deciduos (p.e. la puna).

V.C.5c(1) Con numerosas formas suculentas.

200 V.C.5d *Vegetación de gramíneas amacolladas de cobertura variable con matas.*

V.C.5d(1) Con plantas almohadillas que pueden ser localmente más importantes que las matas (p.e. la puna al sur de Oruro, Bolivia).

V.C.6 VEGETACIÓN GRAMINOIDE BAJA SIN SINUSIA LEÑOSA.

201 V.C.6a *Comunidades de gramíneas bajas.* Su estructura y su composición florística pueden variar debido a la considerable fluctuación de las precipitaciones en los climas semiáridos (p.e. la pradera de gramíneas bajas [*Bouteloua gracilis* y *Buchlöe dactyloides*] del Colorado Oriental).

202 V.C.6b *Comunidades de gramíneas amacolladas* (p.e. las comunidades de *Poa colensio* de Nueva Zelandia, y la puna alpina seca con *Festuca orthophylla* del norte de Chile y Bolivia meridional).

V.C.7 VEGETACIÓN GRAMINOIDE MESOFÍTICA INTERMEDIA A BAJA (PRADO).

203 V.C.7a *Comunidad graminoide cespitosa*, a menudo rica en forbias, por lo general dominada por hemicriptófitas. Ocurre principalmente en altitudes inferiores de clima frío y húmedo en América del Norte y Eurasia. Muchas plantas pueden permanecer verdes, al menos parcialmente, durante el invierno, incluso bajo la nieve en las latitudes superiores.

204 V.C.7b *Prado alpino y subalpino de las latitudes superiores*, en contraste con los tipos de vegetación de páramo y puna de las latitudes inferiores. Por lo general, húmeda gran parte del verano debido al deshielo.

205 V.C.7b(1) Rica en forbias (p.e. en la península Olympic, Wáshington).

206 V.C.7b(2) Rica en matas (p.e. en las Montañas Rocosas del Colorado).

207 V.C.7b(3) Comunidad nival. Formación clara, rica en pequeñas forbias y en matas semejantes a las forbias (p.e. *Salix herbacea*). En las altas latitudes equivale al superpáramo de las bajas latitudes (véase V.C.5b).

208 V.C.7b(4) Prado de alud, que aparece como estrechas bandas de vegetación graminoide entre los bosques en laderas muy empinadas de altas montañas donde los aludes, que descienden anualmente en la primavera, impiden el crecimiento del bosque. Estructura variable; pueden tener algunos arbustos y árboles dañados.

V.C.8 TUNDRA GRAMINOIDE. Como en el caso de la tundra arbustiva (IV.D), el empleo del término "tundra graminoide" debe limitarse a las altas altitudes, des decir, más allá del límite polar de los árboles.

209 V.C.8a *Tundra graminoide amacollada.* La mayor parte de las graminoides crecen en forma amacollada. Entre las macollas existen a menudo musgos y líquenes. (p.e. tundra de *Eriophorum* en Alaska Septentrional).

210 V.C.8a(1) Estacionalmente inundada.

211 V.C.8b *Tundra graminoide cespitosa.* Las graminoides forman un césped más o menos denso, con frecuencia con una diversidad de forbias y de matas (tanto cespitosas como reptantes). Pueden ser comunes los líquenes (p.e. tundra próxima al lago Iliama, Alaska).

212 V.C.8b(1) Estacionalmente inundada.

V.D *VEGETACIÓN DE FORBIAS.* Las comunidades vegetales consisten principalmente en forbias, es decir, la cobertura de éstas es superior al 50%. Puede haber graminoides pero cubren menos del 50%, y a menudo mucho menos.

	V.D.1	COMUNIDAD DE FORBIAS ALTAS. Las formas dominantes de crecimiento de las forbias son de más de 1 metro de altura cuando están plenamente desarrolladas.
213	V.D.1a	*Principalmente forbias perennes con flores y helechos.*
214	V.D.1a(1)	Sustrato salino o húmedo gran parte del año (p.e. los pantanos de *Batis-Salicornia* de Florida y las formaciones de *Heliconia-Calathea* de América Central).
215	V.D.1b	*Matorral denso de helechos,* a veces en grupos casi puros, especialmente en climas húmedos (p.e. *Pteridium aquilinum*).
216	V.D.1c	*Principalmente forbias anuales.*
	V.D.2	COMUNIDAD DE FORBIAS BAJAS, con las formas dominantes de crecimiento de las forbias de menos de 1 metro de altura en pleno desarrollo.
217	V.D.2a	*Principalmente forbias perennes con flores y helechos* (p.e. los prados de *Celmisia* en Nueva Zelandia y los prados de forbias de las Aleutianas, Alaska).
218	V.D.2b	*Principalmente forbias anuales.*
219	V.D.2b(1)	Comunidad de forbias efímeras en las regiones tropicales y subtropicales con precipitaciones muy escasas donde, desde el otoño a la primavera, las nubes humedecen la vegetación y el suelo (p.e. en las colinas costeras del Perú y Chile septentrional). Dominadas por forbias anuales que germinan al principio de la estación nubosa, crecen abundantemente hasta el fin de ésta y dan al paisaje un aspecto verde fresco. Durante la estación seca el aspecto es desértico. Existen siempre geófitas y hemicriptófitas criptogámicas o camaéfitas, que pueden dominar localmente. Algunas fanerófitas pueden aparecer como reliquias de bosques nublosos naturales.
220	V.D.2b(2)	Comunidad de forbias efímera o episódica de regiones áridas: "el desierto florido". Predominan las forbias de crecimiento rápido, a veces concentradas en depresiones donde puede acumularse el agua. En ocasiones estas comunidades pueden formar sinusias en las formaciones de arbustos o matorrales de las regiones áridas [véase IV.C], (p.e. en el desierto de Sonora).
221	V.D.2b(3)	Comunidad episódica de forbias. Comunidad irregular de estructura y composición variables, que se desarrolla en las partes secas de los cauces fluviales durante los estiajes de más de dos meses de duración, o en zonas temporalmente secas del lecho de ríos que tienden a cambiar de cauce con más o menos frecuencia (p.e. en los lechos fluviales de los principales ríos de los llanos de Colombia y Venezuela).
	V.E	*VEGETACIÓN HIDROMÓRFICA DE AGUA DULCE* (vegetación acuática). Como en el caso de la comunidad acuática de manglares, la formación de helófitas no se ha clasificado en esta categoría. Por ejemplo, las formaciones de *Typha* se consideran comunidades graminoides de altura intermedia, húmedas o inundadas la mayor parte del año [véase V.B.5a(1)].
	V.E.1	COMUNIDAD DE AGUA DULCE ARRAIGADA. Compuesta de plantas acuáticas que están sostenidas estructuralmente por el agua, es decir, que no se sostienen por sí solas, en oposición a las helófitas.
222	V.E.1a	*Formación tropical y subtropical de forbias* sin contrastes estacionales apreciables (p.e. las formaciones de *Victoria regia* del Amazonas).
223	V.E.1b	*Formación de forbias de las latitudes medias y altas* con contrastes estacionales importantes (p.e. las formaciones de *Nymphaea* y *Nuphar*).
	V.E.2	COMUNIDAD DE AGUA DULCE QUE FLOTA LIBREMENTE.
224	V.E.2a	*Formación tropical y subtropical de plantas que flotan libremente* (p.e. *Pistia, Eichhornia, Azolla pinnata,* etc.).
225	V.E.2b	*Formación de las latitudes medias y altas que flotan libremente* y que desaparecen durante el invierno (p.e. *Utricularia purpurea, Lemna*).

Cartografía

LOS LÍMITES

Las diversas unidades antes indicadas deben delimitarse en el mapa por un trazo que las separe de las unidades vecinas. En un mapa geológico esto es fácil. Ahora bien, en los mapas climáticos o de vegetación la representación es con frecuencia mucho más difícil ya que las unidades que es necesario representar raramente tienen límites muy netos. Esto puede ser cierto para un bosque o para un pantano, pero a menudo hay una transición continua de una unidad a la vecina. Tipográficamente es muy difícil hacer gradaciones y muchos límites dan una precisión que no existe en la naturaleza.

En una escala de 1/5 000 000 los contornos indicados por los autores que ya han publicado mapas no son a veces lo suficientemente precisos. La utilización de fotografías aéreas, sobre todo tomadas a gran altura, pueden permitir precisar los contornos de formaciones cuyos caracteres botánicos se conocen.

Para escalas mayores es necesario emplear fotografías aéreas. Hace falta naturalmente que el fondo topográfico sea exacto, lo que no occurre en todos sitios.

COLORES EMPLEADOS

Siguiendo las ideas aplicadas desde 1924 por H. Gaussen, los principios teóricos son:

1. La manera de aplicar el color: tinta plana, sombreado con rayas o sombreado con puntos para representar los tipos fisionómicos esenciales: bosques, matorrales y formaciones herbáceas.
2. El color empleado indica las condiciones del medio por síntesis de colores elementales, cada uno de los cuales representa un factor mesológico. Esta síntesis, realizada en 1926, ha dado después buenos resultados en los mapas publicados a escala 1/5 000 000 (Unesco-FAO, Países mediterráneos); a 1/1 000 000 (mapas de la India, Sahara, Madagascar); y a 1/200 000 (vegetación de Francia, África del Norte).

Cuando se tiene la costumbre de gama de colores, de una sola ojeada se aprecian las condiciones generales de la vegetación. Estas condiciones son esencialmente climáticas en las pequeñas escalas (1/5 000 000) pero puede destacarse la importancia de algunos suelos (dunas, manglares, pantanos, lagos).

Más adelante se encuenra una hoja plegable, fuera de texto, con ejemplos de los principales colores.

En los países tropicales, donde las condiciones del medio son muy diferentes en la estación seca y la estación húmeda, se utilizan bandas verticales, alternas, que representan los dos tipos.

Para cada color empleado, se utilizan tres intensidades, por ejemplo para el color azul (B): B_1, muy claro; B_3, sombreado bastante intenso; B_2, tinta plana.

Los colores fundamentales son: rojo = R, naranja = O, amarillo = Y, verde = V, azul = B, violeta = P (púrpura violeta), gris = G y marrón = M (para las turberas).

Por medio de superposiciones pueden crearse numerosos matices.

El color ecológico indicado por los números en negritas de la clasificación resulta de la superposición de colores elementales que representan los varios factores del medio. A la escala de 1/5 000 000 los factores esenciales son la humedad y el calor.

Las escalas de colores es la siguiente de acuerdo con la serie del arco iris.

Humedad

Muy seco: color naranja
Humedad mediana: amarillo
Muy humedo: azul
con todos los intermediarios.

Calor

Muy caliente: rojo
Calor mediano: amarillo
Muy frio: gris
con todos los intermediarios.

Un bosque templado, es decir un poco húmedo y de calor mediano es:
azul claro + amarillo = verde claro.
Un bosque tropical, es decir húmedo y caliente es:
azul + rojo = violeta.
Un desierto caliente, es decir seco y caliente es:
rojo + naranja = rojo anaranjado.
Un desierto frío, es decir seco y frío es:
naranja + gris = gris anaranjado.
Etc., etc.

La ausencia de vegetación puede señalarse de la manera siguiente:
blanco: con calor = puntitos rojos
muy frío = puntitos grises.

La naturaleza del suelo puede intervenir: arena, pantano, manglares, etc., signos especiales pueden representarla.

SIGNOS EMPLEADOS

Los signos se inspiran en los que ha publicado H. Gaussen y que han sido aceptados por Braun-Blanquet. Esos signos se han propuesto para los mapas a gran escala. Cada género tiene un tipo de signo y las diversas especies constituyen variaciones en torno a ese tipo.

En las pequeñas escalas, en una carta fisionómica, no se puede entrar en tantos detalles.

Se han distinguido las formas: árbol frondoso, de gran talla, de talla media, arbustivo. Lo mismo para las resinosas.

Signos complementarios: espinoso, suculento, micrófilo, turberas, etc.

Las plantas caducifoliadas se representan sin relleno negro.

Las plantas perennifolias se representan con relleno negro.

Las sabanas y las formaciones cespitosas tienen signos especiales.

Los pantanos, las aguas dulces y las aguas saladas continentales tienen también sus símbolos.

Anexo sobre la representación de los cultivos

En las representaciones cartográfica y geográfica en general, es fundamental la escala.

Mientras que a la escala de 1/5 000 000 no es posible representar los cultivos a pesar de su gran interés económico, a la escala de 1/1 000 000 y con más razón a escalas mayores, se puede inscribir en el mapa lo esencial de las estadísticas agrícolas sobre fondo blanco. La vegetación potencial de las partes actualmente cultivadas, a menudo difícil de conocer, se indica mediante un cartón anexo al mapa y con colores ecológicos. En el mapa principal se indican los límites de las diversas unidades que se distinguen.

A tales escalas, el mapa es al mismo tiempo botánico y agrícola, lo que aumenta mucho su interés para el público no especializado en botánica. Cabe citar algunos ejemplos: mapas de la India a 1/1 000 000 o de Francia a 1/200 000.

Cuando la escala permite representar los cultivos, cada signo tiene un valor estadístico. Por ejemplo, un signo de viña representa 10 hectáreas, un signo de olivo, 10 000 o 1 000 o 100 árboles según su tamaño, etcétera.

El empleo de letras permite indicar el porcentaje del tipo de utilización del suelo en una división administrativa determinada (distrito). Todo ello depende de la escala, pero es utilizable a 1/1 000 000 y a escalas mayores.

La ventaja del fondo blanco es que se pueden emplear signos en color relacionados con la ecología de la planta representada.

Símbolos de vegetación

	Sempervirentes	Caducifoliados
ÁRBOLES		
Latifoliados		
Aciculifoliados		
De más de 50 metros de altura		
Coníferas del tipo de los pinos		
De tronco abombado		
Sobre termiteros		
Microfoliados		
Espinosos		
Palmeras		
MATORRALES		
Latifoliados		
Matorrales altos		
Matorrales enanos, secos, eventualmente deciduos		
Sobre termiteros		
Microfoliados		
Suculentos		
Bambús		
Matorrales		
Palmeras enanas		
Matorrales enanos		
Matorrales enanos con arbustos almohadillados		
PLANTAS HERBÁCEAS		
Herbáceas latifoliadas perennes		
Latifoliadas efímeras o episódicas		
Helechos		
Líquenes		
Musgos		
Salicornia		
Plantas reptantes		
Plantas acuáticas, flotantes, arraigadas		
Pantanos		
Pantanos turbosos		
Dunas		

SC. 72/XXIV-6/AFS